T0327726

New Media, New Policies

Richard Collins is Research Director of the IPPR Media and Communication Programme and Lecturer in Media and Communications at the London School of Economics and Political Science. He is the author of several studies of media and communications including *Culture Communication and National Identity: The Case of Canadian Television* (University of Toronto Press 1990 and 1994) and *Broadcasting and Audio-Visual Policy in the European Single Market* (John Libbey 1994).

Cristina Murroni is the Principal Researcher of IPPR's Media and Communication Programme. Her publications include *New Issues in Universal Service* (with Richard Collins, IPPR 1995) and 'Audio-visual regulation and policy' (*European Institute for the Media Bulletin* vol. IV no. 12). With a background in social sciences and international economics, she has several years' experience in market research (1991–3) and formerly worked for the independent consultant London Economics (1993–4). She speaks good Portuguese, Spanish and French and is a native speaker of Italian.

New Media, New Policies

*Media and Communications
Strategies for the Future*

Richard Collins and Cristina Murroni

Polity Press

Copyright © Richard Collins and Cristina Murroni 1996

First published in 1996 by Polity Press in association with Blackwell Publishers Ltd.

2 4 6 8 10 9 7 5 3 1

Editorial office:
Polity Press
65 Bridge Street
Cambridge CB2 1UR, UK

Marketing and production:
Blackwell Publishers Ltd
108 Cowley Road
Oxford OX4 1JF, UK

Published in the USA by
Blackwell Publishers Inc.
238 Main Street
Cambridge, MA 02142, USA

ISBN 0–7456–1785–9
ISBN 0–7456–1786–7 (pbk)

A CIP catalogue record for this book is available from the British Library and the Library of Congress.

Typeset in Stemple Garamond 10/12pt
by Wearset, Boldon, Tyne and Wear.
Printed and bound in Great Britain by
Hartnolls Limited, Bodmin, Cornwall
This book is printed on acid-free paper.

Contents

Preface and Acknowledgements

One of the pleasures of prefaces is that they are written when the end is in sight! Only those who have written a book will know the keen pleasure and profound relief that comes with dispatch of disk and hard copy to the publisher. But another pleasure is acknowledging with gratitude the contribution others made to the finished work.

This study was undertaken in the Media and Communication Programme at the Institute for Public Policy Research (IPPR) in London between 1994 and 1996. IPPR is a centre left 'think tank' and an independent charity which was set up in 1988. It undertakes research in the public interest and provides a forum for political and trade union leaders, academic experts and those from business, finance, government and the media. Richard Collins and Anna Coote, Deputy Director of IPPR, established IPPR's Media and Communication Programme in 1994.

We are pleased to thank all those who helped us complete this study. Warm thanks go to the members of the IPPR Media and Communication Programme Advisory Committee. Distinguished academics, trade unionists, consumer and industry experts all generously shared their expertise. The help given by the following greatly enriched our work: Duncan Campbell-Smith, Martin Cave, Jane Clarkson, Barry Cox, Sally Davis, Rupert Gavin, Tony Giddens, Peter Golding, Andrew Graham, Bradley Herrmann, Patricia Hewitt, Patricia Hodgson, Chris Hopson, Justin Jameson, Dominic Loenis, Peter McInerney, Kip Meek, Jeremy Mitchell, Derek Morris, John Newbigin, Chris Powell, Jane Reed, John Sanderson, David Souter, Sandy Walkington, David Ward, Andrew Whyte and Stephen Young.

We also acknowledge Patricia Hewitt's particular contribution to this work. Patricia's support, when Deputy Director of IPPR, continued after her move to Andersen Consulting, notably through her able chairing of the programme's Advisory Committee. Julian Eccles and Mike

Craven helped us extend our contacts in the media and communications industries; without them much less would have been possible. Eric Barendt, Charles Brown, James Curran, Nick Fitzpatrick and Mark Philips all commented on chapter drafts. Their comments, James Curran's and Justin Jameson's discussion papers (published in the IPPR Media and Communication Programme), and the contributions to seminars (which formed part of IPPR's Media and Communication Programme) made by Daniel Alexander, Lord Borrie, Martin Cave, Don Cruickshank, John Enser, Michael Grade, Andrew Graham, Erica Jong, Bill Melody, Claire Milne, Jeremy Mitchell, Graham Murdock, Simon Olswang, Frank Panford, Robert Pinker, Michael Redley, Andrew Sharp, Clare Short, Chris Smith and Bernard Williams, all informed our findings. 3 Com supported the seminars and publication of IPPR's Media and Communication discussion papers.

This is a jointly authored work whose authorship extends beyond those named on the spine. Drafts have been reworked, ideas shared and reshaped, minds changed in the course of discussions within the research team and the Advisory Committee. So much so that for many sections no one can claim authorship. Richard Collins and Cristina Murroni are the principal authors. This means that they know better than anyone how indispensable were Anna Coote, James Purnell and Elena Cappuccio's contributions. We have shamelessly pillaged many entries to our glossary from that Justin Jameson wrote for his discussion paper for the IPPR Media and Communication Programme – thanks, Justin!

Of the project team, Richard Collins and Anna Coote were there from go to whoa. Cristina joined the team in November 1994 and saw the study through to completion. James Purnell joined in February 1994 but in October 1995 was head hunted to join the BBC. Ironic to lose a member of the team which had applauded the BBC's entrepreneurial turn to the BBC itself! His successor, Elena Cappuccio, was an indispensable member of the programme team from November 1995 to February 1996. Anna Coote's contribution demands especial testimony. Without her, IPPR's Media and Communication Programme would not exist. Her skills in project definition and management and in fund raising created the context in which this project became possible. Her writing and editorial work, her nose for humbug and pomposity and her insistence on posing hard questions saved the principal authors from many silly mistakes and relieved the reader from a lot of badly knotted prose. We are also indebted to Sarah Neal and Adam Jacobs for their research on the film industry and regulatory institutions respectively.

Of course, without our funders – British Telecom, the Cable Communication Association, London Weekend Television, Mercury Communications, News International and Pearson – this study would not have been possible. Our research inevitably addressed issues in which our sponsoring firms had substantial material interests and representatives of the funders were members of our project Advisory Committee. It is a pleasure to testify here that they were scrupulous in not seeking to shape our findings other than through the power of their arguments. Their contributions to the lively debates which took place at Advisory Committee meetings were much valued. This book would have been the worse without their expert knowledge and spirited contributions to discussion. We are grateful to everyone who helped us and absolve them here of responsibility for any and all of the mistakes: we have to take responsibility for them!

Richard Collins and Cristina Murroni

Introduction

Media and communications have never been thought more important. Pervasively they are seen to be key agents of change. In the political domain many, who are otherwise sceptical of News International's assertions, give credence to the headline 'It's *The Sun* wot won it!' In economic life, claims that Western economies are becoming post-industrial information societies have become political orthodoxy. And, culturally, mass media and communications are seen to be primary agents in fostering knowledge and understanding, or, when viewed negatively, in the erosion of established and valued identities and cultures.

Politicians no longer ask supporters to look forward to a future with a chicken in every pot but to one which offers a laptop on every lap. Belief in the unique importance of media and communications is manifest, whether stated in the 1995 G7 summit on the Global Information Society, in Tony Blair's Speech to the Labour Party Conference of 1995, in the European Union's Delors and Bangemann Reports (Commission of the European Communities 1993; Bangemann 1994) or most effusively in US Vice-President Al Gore's claim that 'The information infrastructure is to the US economy of the 1990s what transport infrastructure was to the economy of the mid-20th century' (*Financial Times*, 19 September 1994, p. 22).

The Delors White Paper claimed that the media account for 5 per cent of European gross domestic product (GDP) and are one of only three industrial sectors where jobs will grow. The Commission predicts that the audio-visual industry alone will create 2 million European jobs in the next decade. Delors echoes arguments of researchers like Marc Porat (1977) and Daniel Bell (1973) in the USA, Simon Nora and Alain Minc (1978) in France and Stuart Wall (1977) in the UK who have argued that a world economic revolution is taking place as advanced countries shift from industrial to information economies and as an international division of labour and global economic interdependence accelerate.

Efficient communication systems have become increasingly important to what remains of manufacturing in the old industrial heartlands. Just-in-time production and flexible specialization can only work with good communications between suppliers and assemblers, customers and producers. Automobile manufacturers have merged with information technology firms and have even developed their own telecommunications divisions. The banking and finance sectors manipulate symbols of stored value. The entertainment industries manufacture symbols to fill increased leisure time. The growth of world trade has far outpaced the growth of global GDP, and trade in services, very often in symbols, is growing faster than trade in material goods. Telecommunications provides cheap and immediate long distance communication for command and control of an interdependent global economy.

The growing economic importance of media and communications has combined with a deeply rooted faith in technology to conjure up a seductive vision of a new generation of media and communications technologies. The information superhighway is a powerful metaphor which signifies both the seamless integration of all media and communications into a single global bitstream and the culmination of the forces of economic and social modernization in a new technological Utopia. US Vice-President Al Gore put it thus:

> I believe that the creation of a network of networks, transmitting messages and images at the speed of light across every continent, is essential to sustainable development for all the human family. It will bring economic progress, strong democracies, better environmental management, improved health care and a greater sense of shared stewardship of our small planet. (*Financial Times* 19 September 1994 p. 22)

Scepticism in the face of such inflated rhetoric is almost irresistible. But navigating between overblown claims and an excessive scepticism is far from easy. Not least because the left approaches the task with a weighty, and sometimes unhelpful, intellectual inheritance.

The old left

The failure of Marxist predictions of the transformation of capitalist society into socialism led to a search for explanations. Broadly, the ruling class's success in gaining ideological dominance became the left's preferred explanation for its failure. In this vision of things, the mass media were seen as the principal agencies of what Terry Lovell has called 'the production and dissemination of erroneous beliefs whose inadequacies are socially motivated' (1980 p. 51) and which reproduced an inequitable social order. This analysis chimed with twentieth-century political theory's emphasis on the role of ideas, rather than force, in holding societies together. It was elaborated as a general theory by revisionist Marxists, notably by Stuart Hall. Adaptations of the core thesis were articulated by feminists and anti-racists to account for the reproduction of patriarchal and racist social formations.

Hall argued that the media are in a state of deepening crisis that can only be understood as part of a wider crisis of capitalism. However, the media are not just a symptom: they are also a cause of the endurance of a system against the interests of the majority. Mass circulation newspapers are bourgeois papers. Even those aimed at a working-class readership serve to persuade readers to believe in a national or common interest that obscures their fundamental class interests. The working-class press 'do not produce representations which allow class members to *recognise* themselves as working people, as the producers of wealth, as an exploited class, as the producers of surplus value, as a subordinate culture, as active and militant trade unionists, as men and women who have nothing to lose but their chains' (Hall 1978 p. 41).

Similar arguments were articulated in analyses of the media (see, *inter alia*, Glasgow University Media Group 1976; 1980) and in launching media enterprises which challenged capitalist and patriarchal norms in their editorial stance and internal organization. Instances included worker managed newspapers like the *Scottish Daily News* and *News on Sunday* (see McKay and Barr 1976; Chippindale and Horrie 1988) supported by public funds from national government and local Labour councils. The left's aspirations have focused on the power of the state to regulate privately owned media and to establish media owned and controlled by the public to countervail the spurious free press which 'will work to reproduce society in its already given class structured form' (Hall 1978 p. 41).

In spite of the importance attributed to the media in reproducing a dominant ideology hostile to socialism and the labour movement, there have been few official Labour Party policies and publications on media and communications. In 1974 a Labour Party study group chaired by Tony Benn MP and established by the Home Policy Committee published *The People and the Media* (Labour Party 1974) and in 1991 the party issued *Arts and Media: Our Cultural Future* (Labour Party 1991) which in thirty-four pages devoted only two to the media and two to the cultural industries.

Paternalist corporatism

A second framework in left and Labour thinking is rooted in doctrines of public ownership leavened by Reithian[1] notions about media content. From this paternalist corporatist viewpoint, the privately owned press posed the chief problems within a structure of media and communications which, until the 1980s, was characterized by public ownership of the communications infrastructure. The Royal Mail, British Telecom, Hull Telephones, Cable and Wireless, the BBC, and the commercial broadcasting infrastructure owned and run by the Independent Broadcasting Authority were all publicly owned and the commercial broadcasting sector was firmly under the thumb of a regulator able to control entry to the market.

All changed when the publicly owned communications infrastructure began to be privatized in the early 1980s. The old structure of centralized control over broadcasting unravelled in the late 1980s when the IBA's favoured satellite television monopoly, BSB, was put to the sword by News International's Sky Channel. A system dominated by public ownership changed to one dominated by private ownership without the left developing an analysis equal to changed circumstances. Gloomy predictions about the consequences of commercial competition which pervaded Labour's *The People and the Media* (1974) were reiterated in amplified form in spite of the flawed predictions of the document. Notable among them: that by the 1980s there would only be two or three national newspapers; that 'there was no future for commercial radio' (p. 14); that 'public ownership of all transmission facilities' (p. 15) should be maintained, i.e. there should be no private ownership and no competition in cable or telecommunication networks.

Perhaps citations from 1974 appear a little dusty in the mid-1990s. But it is remarkable that, apart from the brief consideration of the media in *Arts and Media* (Labour Party 1991),[2] one has to look so far into the past for a full account of a Labour policy for media and communications. The most extensive and coherent left analysis of the UK media is Curran and Seaton's often reprinted *Power without Responsibility* (1991). Indeed one of the authors, James Curran, was a member of the study group which produced *The People and the Media*. In one of the passages which has survived through the successive editions, the authors state:

> Broadcasting – monopoly or duopoly – always depended on an assumption of commitment to an undivided public good. This lay beneath all official thinking on radio and television until the 1970s. In 1977 the Annan Report abandoned this assumption, and replaced it with a new principle of liberal pluralism. The ideal ceased to be the broad consensus – the middle

ground on which all men of good sense could agree. Rather it became, for Annan and those who supported or inspired him, a free market-place in which balance could be achieved through the competition of a multiplicity of independent voices. The result has been confusion and crisis, from which no new doctrine has yet emerged. (Curran and Seaton 1991 p. 296)[3]

This rejection of private ownership and competition is fundamentally flawed. Publicly owned broadcasting in the UK has been strengthened by competition. Commercial broadcasting has provided a voice for interests and identities to which the public system had been closed.[4] The emergence of British Telecom as one of the strongest communications operators in the world is inseparable from its exposure to competition. Adaptation to new circumstances, discovering and meeting the needs of consumers, innovative products and services and efficient use of resources – all these things are much more likely to occur in a decentralized system of competing firms. Indeed, in the context of an argument for a social market media policy Curran himself states: 'Labour governments have been characterised by a deep conservatism when it comes to media policy. It has been left to Conservative governments to innovate, to break free from conventional thinking and introduce bold new developments that, by and large, have been marked improvements' (1995 p. 9).

New right

In contrast to the silence of the majority of the left, or its reiteration of time honoured arguments, the right has published, and translated into policy, a radical new analysis and – to the left's dismay – a new media and communications regime (see, *inter alia*, O'Malley and Traherne, undated; O'Malley 1994). According to Curran and Seaton: 'the starting point of the free market approach is that consumers are the best judges of what is in their interests' (1991 p. 337). The policy aim is no longer correction of individuals' choices in Reithian fashion but allowing each individual to consume what he or she wants and thereby bring about the allocation of goods that optimizes the collective interest – which is viewed as the sum of individual interests.

In the 1980s, commercial and academic proponents of this approach argued that there was no reason to maintain the broadcasting duopoly, because justification for it, spectrum scarcity, was evaporating. Cable and satellite would allow new entry on an unprecedented scale. Greater competition would maximize consumer welfare and allow an overdue separation between state and broadcasting. Competition would destroy insiders' privileges and foster wider access to resources and to a voice. This new paradigm was forcefully articulated by writers associated with the Institute of Economic Affairs (see, *inter alia*, Brittan 1987; Veljanovski and Bishop 1983; Veljanovski 1987). It found realization in the Peacock Report (1986) and the policy initiatives which followed. It achieved an apotheosis in Rupert Murdoch's seminal McTaggart Lecture in 1989 in which he argued that his satellite television initiatives represented an interim stage 'in a transition in which British television will come of age, reaching maturity by breaking out of its self-imposed duopoly and entering a time when freedom and choice, rather than regulation and scarcity, will be its hallmarks'.

There has been rather less discussion on left and right of telecommunications than of the press and of broadcasting. In 1981 Mercury was licensed to compete against BT and the BT network was opened to commercial service providers. In 1982 BT's privatization was announced and the current telecommunications regime was put in place with the promulgation of the Telecommunications Act 1984. The government's decisions were based on experience in the United States (see, *inter alia*, Society of Telecom Executives 1983 p. 3), where telecommunications had been provided by commercial companies and where in 1982 Judge Greene had set in train the divestiture of AT&T, which was implemented in 1984. The UK rapidly went beyond the US model and became what is now seen as a laboratory in which world class practices and firms have been developed. Indeed, few now argue for renationalization of BT and criticism of the current UK telecommunications regime avers that liberalization has not gone far enough, notably in arguing for a breakup of BT.

However, the practice has not been as consistent as the theory. Margaret Thatcher's enthusiasm for liberal economic measures sat uneasily with her moral authoritarianism and attempts to intimidate broadcasters – whether public or private. More important, liberalization, privatization and competition did not go as neatly hand in hand as had been promised by proponents of change. As Armstrong, Cowan and Vickers argue in their review of the UK experience of regulatory reform: 'Regulation that was intended to be "with a light rein" has had to be supplemented and tightened repeatedly. Effective competition has not developed as initially hoped, and competition policies for the utility industries have had to be strengthened along with regulatory policy' (1994 p. 355).

This failure is understandable. Economists have long recognized that there are types of production and distribution which do not fit the standard paradigms of neo-classical economics. Such cases are known as market failures.

Limits of the market in media and telecommunications

For different reasons, broadcasting, telecommunications and the information sector as a whole were widely recognized to be exceptions of this kind. However, identifying these as sectors where markets do not work in textbook fashion does not establish a watertight case for non-market mechanisms and state intervention. First, technological change can alter the way markets work. Second, an imperfectly functioning market may well be preferable to an administered, interventionist alternative. Case by case judgement is required to assess the balance of advantage.

Telecommunications, or at least the local loop, is usually thought to be a market failure on the ground that service can be provided most cheaply by a single firm (natural monopoly). However, some contend that economies of scope (such as those conferred by integrating electricity and telecommunication supply or the supply of cable television with telecommunication services) may mean that a natural monopoly in telecommunications no longer exists – even in the local loop. But such arguments do not abolish the problem, they only restate it. If economies of scope outweigh economies of scale, then a single firm able to combine economies of both scope and scale might well still provide the most efficient means of delivering a service (for a fuller account of telecommunications market failures see appendix 1).

Broadcasting is usually thought to exemplify a failed market because of the 'public good' character of the broadcasting market. Public goods are goods and services which are indivisible and non-excludable, where consumption by one person does not diminish the possibility of another person consuming. For example, a panoramic view is a public good, whereas an apple is not. My consumption of an apple means the apple is not there for another person to consume, but my seeing a view does not diminish that view and it remains available for others to see. Broadcasting is this class of good: adding an extra listener to a radio programme makes no one worse off and that extra person better off. Welfare is therefore maximized by free access to broadcast signals. The possibility of free access, however, means that all have an incentive to 'free ride' and to enjoy services paid for by others. Hence the development of subscription broadcasting, where access to services is available only to those who pay. But subscription broadcasting involves a welfare loss: some people are worse off than they might otherwise have been, because they are deprived of access to programmes which they could have consumed at no additional cost to themselves or to others. The technical failure of free-to-air broadcasting markets may in fact bring social gains beyond those attainable in subscription broadcasting, even if subscription broadcasting is a technically better functioning market.

The economics of the information sector are also odd because of the unusually large gap between average and marginal costs of production and distribution. The cost of production of an additional copy of a film, video or printed work, the addition of one more viewer or listener to a broadcast, an extra minute on a telephone call, is close to zero. The disparity between first copy and second and subsequent copy costs means that there are potentially very high returns to economies of scale in the media industries. The first copy cost of a newspaper includes the capital costs of the printing works and editorial offices, printers' and journalists' salaries, as well as the cost of the ink and paper used to print the first copy. Second and subsequent copies cost little more than the extra ink and paper needed for their production. So, if producers can extend markets, in time or space or both, profits are likely to increase more than proportionately.

In spite of these market failures, changes to media and telecommunications introduced under the new right have produced some striking results. Telephone penetration has risen from less than 70 per cent of British households prior to liberalization in 1982 to about 93 per cent (with cable companies in some cases providing service to homes not formerly served by BT); service quality has risen from a demonstrably poor international standard (Melody 1990) to levels comparable to the best OECD countries (OECD 1990). True, technological change played a part in these improvements but the organizational re-engineering and cultural change in BT which followed competition had an enormous impact. BT was forced to improve efficiency and the increase in total factor productivity achieved is by far the largest among the privatized industries and took place in the sector where competition grew most vigorously.

However, in spite of increased competition, big firms – public and private – still dominate media and communication markets. After ten years of competition BT had 92 per cent of telephone lines and 87 per cent of the value of business services. The BBC has near 20 per cent of the share of voice of the UK media market (*Financial Times* 21 March 1995 p. 11). The four terrestrial channels have 94 per cent of television viewing. The top four newspaper chains have 88 per cent of national newspaper readership (Oftel 1994b; Barnett 1995; Department of National Heritage 1995). News International alone accounts for 39.2 per cent of national daily newspaper circulation and 38.5 per cent of national Sunday newspaper circulation. News International also has the largest shareholding in the dominant satellite television system, BSkyB. Its control of the market leading encryption system, of transponders on the market leading satellite and of programming rights has led some of its competitors to complain to the Office of Fair Trading and the Competition Directorate of the European Commission (see, *inter alia*, Office of Fair Trade press release no. 50/95, 1 December 1995). The incumbents' dominance and control of key 'bottlenecks' enables them to govern market entry. The regulatory functions of the modern state cannot, yet, wither away.

Technological change and re-regulation have delivered striking benefits. But benefits have been unequally enjoyed because competitive pressures are not evenly spread. In the international and long distance telecommunications markets there is effective competition, but BT's monopoly of access to the residential market is being challenged very slowly. Its main competitors in the local loop, the cable operators, have so far provided about 1.2 million lines (ITC data October 1995) against BT's 26 million. In broadcasting, new services, such as satellite television and commercial national radio, have provided new consumption opportunities and have encouraged incumbents to change their programming to reduce their losses of viewing share. In both domains incumbents' ability to roll out technologically advanced infrastructures endangers competition and threatens to exclude new entrants – e.g. BSkyB's control of gateway facilities in analogue, and possibly digital, subscription markets.

Universal service at affordable cost is under threat in broadcasting and telecoms. Some programming – notably sports programming – hitherto available to viewers at zero cost is now available only to those who pay. The exclusion of viewers from encrypted programmes, to which they could have access at zero incremental cost, in order to levy charges on subscribers, raises an increasingly urgent policy question. New technologies such as digital broadcasting, left to develop in a regime of unregulated competition, are likely to deliver monopolist gatekeepers or consumer-unfriendly and incompatible rival standards. Falling costs of transmission, distribution and reproduction of information could bring us free universal access to information of all kinds – from person to person communication to entertainment, from stored knowledge and reference material to the enhanced ability to work by using the 'know-how' embedded in computer software. Notable examples include Internet, where the cost of usage is not distance related, and computer shareware. However, controls over intellectual property are tightening. Copyright protection has been extended far beyond that necessary to provide incentives for the production of new works. The seventy years of protection now extended *after* the death of the author provide owners of copyright, few of whom will actually be authors, with an extended period for extraction of monopoly rents from users. The technological potential of cheap transmission and reproduction is being compromised to maximize returns to the owners of intellectual property.

Some on the right continue to believe in the simplistic equation of increasing competition and withering regulation. A representative of one of the leading new right 'think tanks', summed up those hopes recently when he said that he 'would like to see all of [the regulators] promote competition to such an extent that their Offices wither away' (Robinson 1993). This is neither possible nor desirable.

In broadcasting and the information sector market failure, for example non-excludability, has socially desirable consequences, which public policy should seek to realize not curtail. In telecoms there is evidence to suggest that, in spite of technological change, some segments of the sector exhibit pronounced economies of scale and scope. If that is the case economic theory suggests that resources can be used most economically by allowing firms to exploit these economies. It is therefore possible, and some would claim likely, that an end result of competition in telecommunications would be a return to monopoly. Would such an outcome be in the public interest? Our view is that competition – even that which is artificially created by regulation (e.g. through entry assistance regimes or other devices of 'asymmetrical' regulation) – is likely to keep dominant firms more efficient than they would otherwise be.

New synthesis

The assumptions of the new right and the old left are fundamentally opposed. The new right starts from consumer sovereignty, the old left from a desire to protect the public from itself – correcting consumer tastes, or at least the results of consumer choice. The new right's solution, competition, is the old left's problem. The new right's problem, corporatism, is the old left's solution. In these circumstances, it is not surprising that the media reform debate has resembled a dialogue of the deaf. This would have been of limited concern if one of the sides had proved its case. However, media policy on both sides has been restricted by different, but equally flawed assumptions. We need to abandon such tribalism to allow a new, radical, synthesis of these approaches.

The old left theory of the audience was flawed. As Anthony Smith (1978) points out, the mass audience, far from being automatons under the control of the media, are closer to tyrants to which producers have to bow. For Russ Neuman (1991) the individual cannot be both the gullible tool of press propaganda and the stubborn resister of 'improving' public service programmes. This error is aggravated by the old left's distrust of commercial competition. By meeting consumer needs more efficiently, competition can improve overall consumer welfare. The evidence refutes pessimistic predictions like those made in *The People and the Media*.

Rather than the *reduction* in national newspaper titles which the Labour Party foresaw, the number of national newspapers has *increased*, as has the overall number of media service providers. In 1974 ITV was the only commercial television channel in Europe and the BBC was the only public service broadcaster in the UK. Now Europe has 217 commercial channels (*Screen Digest* January 1996 p. 9), often transmitted from offshore locations by satellite. Channel 4 has demonstrated that there is more than one way to be a public service broadcaster. Far from having no future, the commercial radio sector has expanded both in the number of stations and in the amount of advertising, thereby meeting consumer needs that were previously ignored. In telecommunications, quality of service and productivity have improved beyond recognition since the privatization of BT and the introduction of rival commercial competitors. In 1974 the UK had two publicly owned telecommunication service providers (and Cable and Wireless serving overseas markets). Now there are more than 150 licensed companies, only one of which, Kingston Communications, formerly Hull Telephones, is publicly owned. The combination of a paternalist view of consumers and a pessimistic view of competition led the old left to cede competition to the new right, when they could have been putting the market to work for traditional socialist ends.

The errors of the new right stem from an overly simplistic equation of changing technology and growing competition. Technological changes may make a competitive market more feasible, but of themselves do not guarantee effective competition. After seventeen years of liberalization and privatization, media markets remain clogged by monopoly.

The left is justified in stressing the structural inequalities of UK media and communications. The remedies of the new right will not end them. Competition and de-regulation do not go hand in hand in the media. Certainly, a promise of technological change is not sufficient to justify the end of media and communications regulation.

If we need regulation, what kind should it be? The European Convention on Human Rights explicitly confers rights to 'hold opinions and to receive and impart information and ideas without interference by public authority and regardless of frontiers' (Article 10.1). What role is there for government here? Some claim that specific media and communication regulation is no longer required as technological change has opened markets. A stronger UK competition law to include Articles 85–90 from the Treaty of Rome and a muscular competition authority would sort out many of the problems. Or so the argument runs. In fact neither freedom of expression nor effective competition can be secured without a framework of law which defines and enforces rights. Otherwise, as Isaiah Berlin remarked, 'freedom for the pike is death for the minnows' (1969 p. 124). A stronger competition policy, though desirable, is not sufficient to secure the public interest in media and communications. For the economic characteristics and political importance of media and communications are different from those of steel and shoes. But their distinctive character should not be an alibi for hobbling a sector which promises so much in terms of jobs and wealth creation.

Liberalization has created a large number of innovatory new firms, from independent producers in broadcasting to the scores of new licensed telecommunication service providers; it has also revivified incumbents. Yet, in spite of technological change, media and communication markets do not fit the paradigms on which competition law is built. In telecommunications the dominance of incumbent firms is both inherited from the past and sustained by economies of scale. In free-to-air broadcasting signalling of preferences by consumers to producers is poor; encrypted broadcasting and protection of authors' rights promise to maximize returns to suppliers rather than to further the interests of citizens and consumers. What is needed is competition policy where competition can thrive in the public interest and regulation where it cannot.

The UK is better placed than any other country in Europe to benefit from growth and integration of global media and communication markets. Works in English circulate more easily than do works in other languages. And the large size and revenue base[5] of the English language mean that there are high potential returns and, crucially, that costs of expensive and high quality works can be amortized in the English language market before being exported. This advantage is one of the reasons behind the success of the US film industry.[6] Moreover, English language producers benefit from the status of English as the world's preferred second language. It is easier to sell an English language work into the German and Spanish language markets than it is to sell a Spanish work into the German and English markets.[7]

Of course, this does not mean that English language producers will always succeed, but there are undoubtedly opportunities for English language products and producers which do not exist for other language communities. It is the *Wall Street Journal* and the *Financial Times* which are competing to be the global newspaper for political and financial elites and not *Les Echos*, the *Frankfurter Allgemeiner Zeitung* or the *Asahi Shimbun*. More than all its partners in the European Union, the UK has an economic interest in international trade in information products.

Criteria for regulation

Media and communications policy should promote, not simply defend, the public interest. Positive objectives, other than economic aims, should be stated and used as criteria to assess policy and regulation.

UK policy makers lack explicit statements of fundamental social criteria. In other countries the constitution will offer a clear reference to national values and a first guidance when conflicting values have to be fulfilled. For example, US policy makers can refer to the fundamental values of 'life, liberty and the pursuit of happiness'. When they need to strike a balance between liberty and order, they are guided by the knowledge that freedom of expression is a constitutional right guaranteed in the First Amendment. British policy makers have no such guide. They also lack a well established repertoire of legal and regulatory precedents: the UK has never been compelled to develop general regulatory precepts.

IPPR's Commission on Social Justice was established to develop objectives and strategies for long term policy making in the UK. The Commission proposed a core of ideas that help in understanding the aims of social justice. The key statement is that 'the foundation of a free society is the equal worth of all citizens' (IPPR 1993 p. 3). It argues that the conditions to meet basic human needs are a right of citizenship and that self-respect and personal autonomy depend on the widest possible range of opportunities. Starting with the notion of equal worth of all citizens and proceeding to the radical conclusion that not all inequalities are necessarily unjust, the Commission identified 'four central objectives for government policy': security, opportunity, democracy and fairness (1993 pp. 4–5):

- The objective of *security* builds upon the notion that social justice demands relief from poverty. But the real policy aim is to prevent poverty where possible and relieve it where necessary.
- The objective of *opportunity* points towards policies designed to increase autonomy and life chances. 'What government can do for people is limited, but there are no limits to what people can be enabled to achieve for themselves' (1993 p. 5). In a media context, opportunities are maximized by ensuring free access to information and affordable universal service in media and telecommunications.
- The objective of *democracy* points towards policies designed to ensure diffusion of power within government and between government and people. Accountability of service providers, ensuring that distinct interest groups and regions are adequately served and that the media are democratically controlled, are practical implementations of this

principle in media. Provision of consumer representation (Potter 1988) is also part of this policy objective.

- The objective of *fairness* points towards policies designed to reduce unjustified inequality (or, in the rhetoric of the European Union, 'social exclusion'). It also points to policies aimed at guaranteeing a right to redress (Potter 1988).

The Commission's criteria provide benchmarks through which policy makers as well as the public can judge policy options. They do not tell us by themselves what public policy should be. These criteria provide comprehensive policy objectives, which promote both the public interest at large and the specific interest of consumers of media and communications. Specific goals of media policies, which bring consumer perspectives[8] into the fore, form a subset of IPPR's more general principles.

Potter (1988) and Sargant (1993)[9] and the *Report of the Committee on Financing the BBC* (Peacock 1986), commissioned by the Conservative government, focused attention on the consumer interest in media and communications (see also Mitchell and Blumler 1994). Peacock's advocacy of consumer interests has led some on the left to contest the relevance of the concept of the consumer and to argue instead for the category of citizen as the key point of reference for policy makers. But this is an unhelpful false choice. People are citizens *and* consumers and their interests as citizens and consumers are interdependent.

Armed with these clear policy goals, we can now explore the issues that are making communications policy ripe for change. There are two main challenges. First, how to draw the boundaries between what is necessarily public and what is legitimately private. The distinction between the legitimately private and the desirably public, always difficult and controversial, is now harder to make. Media which formerly were unequivocally public, such as broadcasting, are converging with those which were unequivocally private, like telephony. Is video-on-demand a private or a public medium? Is a computer bulletin board a medium of public exchange or a form of private communication? Technological change has created new media which do not fit comfortably into the established media order and there are striking anomalies in the combination of tightly regulated 'old' media and barely regulated new media. Decisions about invasion of privacy hang on this distinction. So too do decisions about restriction of representation of possibly offensive material – showing violence, sexual activity and sexual difference.

Second, media and communication regulation faces new difficulties of jurisdictional definition and enforcement. One might say that until recent times the media and communication industries in the UK were economically 'innocent'. They were not subjected to significant competition from overseas and they were not considered to have a central role in national economic life. They have now lost their innocence. Technological change has inescapably internationalized media and communications. Television, radio and even newspaper publishing are now international businesses. The end of national control of media and communications can mean either a positive emancipation from the tutelage of the state or a dangerous slide towards anarchy. Radio Atlantic 252 from Ireland, hundreds of satellite television channels, and countless (voice, data and graphics) services accessed over the telephone and Internet are part of the increasing volume of communication services outside the jurisdiction and effective control of UK regulators.

National governments can no longer control all aspects of their media and communications regimes – for which libertarians are usually thankful! A growing number of issues require international and European jurisdiction. But the profound international impact of the model of liberalization developed in the UK laboratory, exemplified in the new European Union telecommunications regime and elsewhere, demonstrates how the UK can set the pace and direction of international change. It could do so again by introducing a wholesale review and reconstruction of communications and media regulation. These criteria and the economic criteria highlighted above should provide a basis for a new media and communication policy for the UK. In the UK, criteria which meet with general approval are not embodied in regulators' remits. For example, the Independent Television Commission has licensed a fifth television channel on the basis of criteria that favoured established media businesses rather than new entrants. At the same time, the government proposed legislation to control dominant, established business and check media concentration. Lack of co-ordination also leads to regulation of cable companies absurdly split between Oftel and the Independent Television Commission – with confusing differences in franchise terms depending on the date franchises were issued. UK media regulation cries out for reform.

Overview

Throughout the book we discuss media and communication policy as a single field, in the belief that distinctions which served us well in the past are no longer appropriate or sustainable. In chapter 1 we consider how the basic infrastructure for wired and wireless communication should be provided, organized and allocated. In chapter 2 we show that the terms on which telecommunication networks interconnect, a seemingly arcane issue, exemplify a general problem of growing importance: how third parties can secure access to essential facilities. There is no dilemma of markets or regulation. The question is how to strike the right balance between markets and regulation.

For example, we propose using markets for allocation of the radio frequency spectrum. However, we do not want market forces to price some spectrum users off the spectrum. We would also like to preserve the government's power to use spectrum to strike bargains in the public interest. Rather than invalidate the use of markets, these objections point to the need for more than one market. Creative combinations of market and administered systems of allocation – a convergence of liberalization and regulation – are, in our view, likely to provide the best basis for securing the public interest in media and communications in the future.

The untrammelled operation of markets in media and communications is dangerous. The left has concentrated its attention on the potential dangers of size and the related question of concentration of ownership. Ownership is a problem because ownership often means control. Control of a large proportion of the media by a single interest, able to influence media content, is intolerable for democracy. Control, not ownership, is the key issue. In chapter 3 we propose that limiting concentration of ownership should be complemented by strengthening institutional, journalistic and editorial independence and thereby weakening the link between ownership and control in media.

Convergence, whether regulatory or technological, means that past values and practices have to be rethought. In chapter 4 we analyse the goal of universal service in broadcasting and telecommunications. Hitherto the universal service obligation (USO) has been a cardinal principle of carriage regulation: post and telecommunication services should be accessible to all at constant prices. But should the content of universal service change as new services come on stream? Should the USO be defined solely as access to the plain old telephone service (POTS), or should it include pretty amazing new services (PANS)? Is there a universal service obligation in media content?

Next, we examine the role of the government in the communications sector. Chapter 5 explores government's ability to influence media content. We argue for complete re-regulation of media content, where freedom of expression is the rule and restrictions to it must be exceptional. In chapter 6 we turn to the audio-visual industry. Should the government grant it special support? How? Are European trade restrictions in the public interest? The most important instance of government intervention in the audio-visual sector, the BBC, is the subject of chapter 7.

Finally, we argue that public policy goals for media and communications can only be achieved through fundamental reform of regulation. Too many regulators give us the worst of all worlds: regulatory remits overlap and leave gaps uncovered. In chapter 8 we propose replacing existing regulatory agencies with a single statutory regulator, Ofcom, working with self-regulatory bodies and controlled by members accountable to the public and independent of government.

1 Market Forces in Telecommunications

Competition is a relatively new concept in telecommunications. The provision of telecommunications services in each national market has been traditionally entrusted to a large, vertically integrated monopolistic operator. Telecommunications were considered a vital infrastructural sector of the economy and a natural monopoly. Most countries believed that the best way to provide telecom services was to have a single operator. Two dominant patterns of ownership and control obtained: in Europe the monopoly operator was publicly owned, and in North America it was privately owned, and regulated in the public interest. In the early 1980s, advances in technology, which revolutionized the cost structure of the industry and introduced new IT intensive services, induced a general rethinking of how the telecommunications sector should be run.

Then, the UK and the USA pioneered new ways of managing the telecommunications industry. In the USA in 1982 with the Consent Decree, AT&T agreed to divest its local operating companies (now independent companies often known as the 'Baby Bells') in exchange for a looser regulatory regime which permitted it to enter the new markets emerging from IT and telecoms convergence. In the UK the British Telecommunications Act 1981 broke 'the State monopoly in telecommunications' (Department of Trade and Industry 1982 p. 1). British Telecom was privatized, a national competitor, Mercury Communications, was licensed and end-user terminal equipment was liberalized.[1] Once the model of a single state monopoly had been abandoned, the question of how to ensure that the public interest did not go by default was unavoidable. The UK borrowed a North American solution and established an independent regulatory body for telecommunications – Oftel – to ensure 'fair competition and fair prices' (1982 p. 2). This 'Anglo-Saxon' model – of a liberalized sector made up of competing commercial firms regulated in the public interest by an independent public body – has become a central part of what Eli Noam characterized as a 'paradigm shift in the concept of public telecommunications' (1987 p. 30).

At first, competition developed in long distance and international services, but new entrants remained dependent on the dominant operators for access to the local loop, the network that reaches the individual customers from their local exchange. No operator had any commercial incentive to duplicate the costly local network. But the commercial and technological characteristics of telecommunications markets kept changing. A generic type of licence was established for value added network service (VANS) operators, as they were then known. An increasing number of services potentially available on-line, together with the wrangles over most interconnection agreements, sparked operators' interest in reaching customers directly. Further operators – including Energis, Colt, MSF, the cable companies and others – were licensed after the Mercury/BT duopoly was reviewed in 1991. Limited competition has now developed in the local loop, where mobile and cable operators are eroding BT's local monopoly. By 1996 more than 150 operators, including AT&T, had received licences from Oftel to provide telecommunication carriage in the UK.

The experience of the pioneer countries showed that competitive pressures can force telecommunications prices down, while improving the quality and range of services on offer, albeit the benefits accruing from price reduction may be unevenly distributed. In 1987 the European Union also proposed to liberalize telecommunications (Green Paper on Telecommunication, Commission of the European Communities 1987). New services and end-user terminal equipment (telephone sets, answering and fax machines, and so on) were liberalized first, with all services to follow gradually until complete opening of the markets by 1998, except in some of the European Union's poorer states which were permitted to delay liberalization.

In a liberalized telecommunications environment the regulator invariably has to contend with the peculiar problems posed by the inheritance from the past. The regulator does not inherit a 'level playing field' on which new firms of equal power compete freely, but a playing field on which one of the players – the old incumbent – is larger, richer and in almost all respects more powerful than the others. In the media sector, regulation is mainly concerned with ensuring plurality of voices and diversity of content. In telecommunications, one of the regulator's main tasks is to curb the power of the ex-monopolist to allow other firms to operate. The British Conservative government believes the regulator should establish conditions for the 'withering away' of regulation when effective competition has ensured that the telecommunications market no longer needs special treatment. Whether or not it is realistic to expect this Arcadian state ever to be reached is debatable; perhaps it is no more likely to be achieved than the 'withering away' of the state foreseen by Marx. The core of the 'paradigm shift in public telecommunications' is the replacement of an integrated hierarchical network controlled by a single operator with a loose, interconnected web of networks run by different operators with no central control. However, the public interest in ensuring that any telephone can be used to call any other telephone, which could be taken for granted in conditions of monopoly, can only be secured in conditions of competition through regulatory action. For the benefits of interconnection and interoperability of networks are enjoyed unequally by interconnecting operators. To take a hypothetical example: if an operator with two subscribers interconnects with another operator's network with ten subscribers, all subscribers benefit. All are able to call and be called by more subscribers than before. But the operator with two subscribers benefits five times as much as does the operator with ten subscribers. Accordingly, the incentives for dominant operators to offer interconnection are considerably lower than are those for new entrants. And to ensure that the public interest in 'any to any' telecommunications is realized, regulators almost invariably are required to enforce interconnection and set the terms on which it is achieved.

Thus far, competition in telecommunications has been heralded as an end in itself. There are plausible reasons for this. Consumers are less subject to the exercise of power by a single supplier, and even limited competition is likely to give competing suppliers greater incentives to respond to consumer demands than is the case under monopoly. However, competition can only be a means to an end. The number of licensed operators in the UK telecom market, a condition unique in Europe, cannot be considered as evidence either that competition is thriving or that social goals are being achieved.

Competition has delivered many benefits, but at the cost of increased controls and regulation. And these benefits are not experienced universally. Accordingly, regulation is required to ensure that the UK is endowed with a modern and efficient infrastructure and that all users enjoy greater choice and better value for money. Thus far, Oftel's principal concern has, rightly, been to reduce the threat to effective competition posed by a dominant operator. As Michael Waterson observes: 'For as long as one operator needs to use another one's facilities differentially from the reverse process (as long as one is dominant), their interaction will need to be managed to keep it reasonably fair' (Corry, Souter and Waterson 1994 p. 108). But from now on telecommunications regulation in the UK must shift its focus. It must ensure that all customers of all service providers have their entitlements as consumers adequately protected and ensure that customers are able to obtain redress for any service provider's default in the satisfactory delivery of consumers' entitlements. Regulators must ensure that the telecommunications sector thrives successfully and that citizens can enjoy its full benefits because this is a key industrial sector:

- The ability to communicate cheaply and effectively is a basic right of citizens if they are to fully participate in society.
- The convergence of telecommunications, IT and media has made this sector increasingly important as a means of obtaining and delivering information. The ability to join the new information society is critically dependent on the telecom operators keeping abreast with innovations.
- Telecommunications services are a fundamental input to many service industries. Their availability and price influence production in many other sectors and are a fundamental element in the choice of location for many multinational firms.
- An efficient telecommunication sector is a crucial element of a country's defence system.

Infrastructure versus service competition

Telecommunications has been traditionally considered to be a natural monopoly, that is to say an industry where a single firm is able to produce more output, for the same level of input, than a collection of firms. But evidence suggests that there is no longer a natural monopoly in the trunk segment of telecommunications. The natural monopoly argument is increasingly challenged in other segments as firms exploit technological change (e.g. new transmission methods such as microwave), economies of scope (stringing cables carrying telecommunication traffic along electricity transmission networks) and a host of other factors to provide services rivalling those of incumbent telecom companies. Accordingly, policy makers have turned to the question of competition in telecommunications and its consequences. Here it is worth observing that the case for competition does not rest solely on the contention that telecoms is no longer a natural monopoly. Indeed, in spite of the remarkable successes which have attended the introduction of competition into some telecom markets (business, long distance) the natural monopoly argument is by no means disproved in other segments (the wired local loop).

However, even if in theory a single operator could supply services more economically than several competing suppliers such a theoretical potentiality may not be realized in practice. Monopolies may become lazy, those who run them may do so in the interests of insiders rather than customers, they may choose technologies unwisely and make poor investment decisions, and so on. Thus, competition between several suppliers may, in practice, be preferable to a monopoly because competition forces firms to reduce their margins and improve the quality of their services. And, crucially, because competition is an effective way to ensure that firms are dynamic and innovative.

Competition raises several knotty regulatory problems, and that of the terms on which systems interconnect is among the most difficult. Moreover, there is more than one model of competition in telecommunications. In the USA 'common carrier' agreements, whereby owner(s) of parts of the public network were required to offer access to their infrastructure to other service operators, were an integral part of liberalization. This presumes that parts of the network, notably the local network, are natural monopolies and/or bottlenecks and should be regulated as monopolies or quasi-monopolies. Competition is principally located in services and long distance carriage rather than in network provision. Many argue that such arrangements give weak incentives to modernization or to investment in networks. And the core tasks of the regulator are ensuring that all service providers can have access to bottleneck facilities and that prices to customers, whether final users or firms needing access to bottleneck facilities, are fair.

The UK, by contrast, has opted for network competition with several operators building infrastructure. In such a regime the regulator must ensure that interconnection between competing infrastructures takes place so that the 'any to any' principle is maintained. But network operators will try to prevent competitors from interconnecting to the most profitable services, will try to charge high prices for interconnection and will try to develop proprietary technical standards (that is, technical features which accentuate the unique and non-compatible attributes of their networks and are legally protected). Under this scenario, in contrast to the regime adopted in the USA, innovation in services is likely to be slower but innovation and investment in networks faster and fuller.

While both types of competition ultimately result in greater efficiency and lower prices, there are a number of clear benefits which derive from the UK approach, which encourages the emergence of rival networks as opposed to stimulating the supply of competing services along the existing network. Following Baer (1994), a number of advantages of infrastructure competition can be identified:

1 It poses a greater threat for the dominant player.
2 It stimulates innovation.
3 It encourages investment in a high return sector, which is often in-accessible to private capital.
4 It results in a cheaper market for telecoms equipment, by shaking cosy relations between suppliers and a monopsonistic buyer.

These benefits must be set against the cost of infrastructure duplication which would happen to a much lesser extent under a single operator, common carrier system. Duplication of parts of the network is a wasteful activity for society as a whole and therefore the gains of greater competition must be traded off against this economic inefficiency. Until recently, duplication was very expensive and the likely benefits of competition were considered to be minimal. However, technological progress, economies of scope, greater volume of traffic and experience of the poor service provided by many monopoly telecommunications service providers have reduced the cost of duplication at every level of the network and increased the level of consumer and policy maker tolerance of the costs of inefficient duplication of provision. Competition has improved overall efficiency in UK telecoms and few now argue in favour of monopoly network provision. But has competition delivered all the benefits claimed for it?

Threat to the dominant player

If network competition poses a greater threat to the dominant player than service competition, we would expect the incumbent to show a better performance and to lose market share more rapidly. There is no doubt that competition has caused the UK incumbent, British Telecom, to undertake a massive programme of efficiency improvement. Total factor productivity, a measure of efficiency improvement in production, increased by 2.2 per cent in 1983–8 and by 7.2 per cent in 1989–94 (London Economics 1994b). This is by far the largest growth among the privatized industries, although it is impossible to tell how much of this efficiency improvement is endogenous (due to privatization) and how much is exogenous (technological change).

BT has also shown remarkable improvements in the quality of services available. Melody commented that: 'prior to privatisation in the UK, British Telecom served only about 60 per cent of British households; it made small villages pay high fees for a public telephone to ensure they were not subsidised; it provided demonstrably poor service by international standards; and it was not up to date in applying new technologies to improve efficiency and expand service offerings' (1990 p. 9). The latest comparative data available on quality of service in OECD countries (OECD 1993) show that BT's performance has improved considerably following introduction of competition. Using call failure rates as a measure of service quality, the UK rates fourth (after semi-liberalized Canada and monopolistic Switzerland and France) among the OECD countries (1993). BT also rates among the top five OECD countries[2] for the percentage of faults cleared within 24 hours[3] and digitalization of both exchange lines and trunk lines.[4] Again, technological change has played a major part in this improvement in performance,[5] but the re-engineering of BT and the enormous cultural change which followed privatization as the company shifted from being a supply driven organization with staff employed on Civil Service contracts to a customer and service oriented private company also had an enormous impact. Indeed this change went hand in hand with and facilitated BT's rapid modernization of its infrastructure.

BT is still dominant among fixed telecommunication operators in the UK and has a huge lead over its competitors. In 1994 – after ten years of competition – it had 92 per cent of the market in volume terms[6] and 87 per cent in value. Its main competitors in the local loop, cable operators, have so far provided over 1.2 million lines (ITC data October 1995), against BT's 26 million. In spite of the new entrants' attractive pricing strategies, such as free off-peak calls to other cable subscribers and lower charges than BT,[7] BT has remained overwhelmingly dominant. However, entry by new players, like Energis and Ionica (which will target the residential market), together with number portability, may further stimulate competition at all levels of the UK telecom market and erode BT's dominance.

Sweden is one of the few EU countries which is as liberal as the UK in telecommunications. Compared to the Swedish incumbent, Telia, BT is more dominant. Telia rates 8/9 whereas BT rates 9/9 on dominant operator's strength, a measure developed 'in terms of market dominance, reputation among major users, pricing, range of services, service quality, range of contracts ... It takes into account areas such as access to new technology ... [the threat of] strategic alliances and global venturing, and (last but not least) the sympathy – or otherwise – with which it is treated by the national government' (Public Network Europe 1995 pp. 40, 117, 123).

BT's dominance is impressive. If competition is the goal, then there appear still to be good grounds for continuing with asymmetrical regulation and entry assistance in the UK to foster competition in telecommunications.

Innovation

Many new services have emerged as new telecommunication technologies have been introduced. The UK compares well to other European countries in deployment of digital systems and services. Call diversion, call barring and call waiting are all available as intelligent networks and the digitalization of switches and lines takes place. Other services are specific to the operator who introduces them, for example BT's caller return and display service, and the differentiation of telephone rings for each call receiver in the household offered by some cable operators. Most operators state that they would not invest their resources in developing such services if they were required to allow others to offer them along their networks. But, generally, competition in the UK has been of a 'me too' kind. BT has remained very much the trend and price setter and Oftel has acknowledged that the range of services on offer in the UK is narrow in comparison to the USA where service competition has been given greater emphasis.

Investments

The UK telecommunication system has seen a number of rival firms and facilities emerge introducing competition in different parts of the UK network and using different technologies, as shown in table 1. These operators have invested billions of pounds to establish rival networks and the UK now has a world class telecommunication infrastructure. In 1994 the Department of Trade and Industry published a commissioned study of the competitiveness of the UK telecommunications infrastructure which compared the UK with other OECD countries. The study

Table 1

Operator	Technology	Competing networks
BT	Fixed wired	Local loop, trunk and international
Mercury	Fixed wired	Mostly trunk and international
Cable companies, COLT, MFS	Fixed wired	Local loop in premium locations
Energis	Fixed wired	Local loop, trunk
Mobile operators	Cellular radio	Local loop
Ionica, Omnicom	Fixed radio	Local loop, trunk

concluded that: 'UK investment levels per subscriber are similar to the USA and lower than Germany and Japan in absolute value, but after account is taken of the differences in the costs of equipment between the countries, the UK obtains more from its investment than either Germany or Japan' (1994 appendix). Although straightforward comparisons are complex, the DTI reported that the UK and the US pay the prevailing international price, while operators in countries like France and Germany, which have less competitive markets, pay as much as three or even four times the international price for their telecommunication equipment. This finding suggests that competition has effectively helped the UK network and services sector improve its bargaining position with the upstream sector of telecom equipment manufacturers.

Competition policy in telecommunications

The Telecommunications Act 1984 charges the Director-General of Telecommunications with fostering competition. What instruments are available for doing so? Are they adequate? Will competition alone create a successful sector and satisfy all relevant stakeholders?

UK telecom regulators have two powerful tools. First, there is the price cap regime, which transfers the power to set prices from the dominant firm to the regulator. The price cap, currently set at retail price index (RPI) minus 7.5 per cent, requires regulated firms to reduce prices and therefore to reduce costs and/or profits. It provides a powerful incentive to maximize efficiency and ensures that some, but not necessarily all, efficiency gains are passed to customers. The price cap regime has been successful in delivering lower prices and is widely considered to be easier and cheaper to implement than rate of return regulation – the main alternative methodology of price regulation, which is used in America. However, price caps have many shortcomings. Final products are also inputs to other products and common overheads are attributed to market segments exposed to competition as well as to areas in which the regulated firm is dominant. As a consequence, it is very hard to determine the extent to which:

- prices reflect costs
- prices of services, in which the regulated firm is exposed to competition, are fair and equitable
- monopoly services cross-subsidize services exposed to competition.

In other words, price cap regulation is not effective in redressing monopoly power. Furthermore, when the level of the price cap is set high, benefiting consumers by passing a high proportion of efficiency gains to them, market entry by new firms is discouraged. Because new firms must offer services at prices lower than the regulated incumbent, capping the incumbent's prices at a level where they are close to cost may preclude entry by new firms.

Second, the regulator may set the conditions on which firms are licensed to supply telecommunication services and stop behaviour which breaches licence conditions. However, the ability to determine conditions of licence is also an inadequate instrument of regulation. Anti-competitive behaviour which is not related to price movements and is not foreseen and provided against in the licence lies outside the regulator's jurisdiction. Moreover, monitoring abusive behaviour and setting appropriate conditions of licence when there are many firms (and more than 150 hold UK telecommunication licences) is an impossible task for anything other than a giant regulatory bureaucracy. The Telecommunications Act provides no powers for the Director-General of Telecommunications to fine firms or require offending firms to pay compensation for behaviour which breaches licence conditions. These inadequacies have been widely criticized. The specific concerns of the telecommunications sector resonate with the general call for changes to the entire regime of competition policy in the UK. The government's 1992 Green Paper on *Abuse of Market Power* (Department of Trade and Industry 1992) favoured maintenance of the present regime and the government has been consistently, and rightly, criticized for its refusal of reform (see, *inter alia*, *Financial Times* 23 November 1995).

A comprehensive discussion of competition policy goes beyond our remit. But some general observations are required to put our sector-specific proposals into context. The current UK regime has serious disadvantages and is notably lacking in deterrents to anti-competitive behaviour. However, because telecommunications is changing rapidly and services and networks are converging with other sectors – notably broadcasting and computing – all possible types of abusive and anti-competitive behaviour cannot be foreseen and provided against in specific licence conditions. The 'administrative' competition regime of the UK contrasts with the 'prohibition' regime specified in Articles 85–90 of the Treaty of Rome to which UK firms are subject in cases of intra-Community trade across frontiers. A prohibition regime, like that of the EU, stresses deterrence rather than punitive action (although the EU Competition Directorate disposes of powerful sanctions) and competition authorities have wider investigative powers as well as the authority to impose fines.

Oftel is presently (1996) seeking wider powers to discourage and penalize anti-competitive behaviour. It proposes a general prohibition of anti-competitive behaviour to replace the detailed specification of conditions in licences. British Telecom has expressed concern that, should Oftel's powers be so extended, one person, the Director-General of Telecommunications (DG), will enjoy enormous discretionary power. Instead BT has proposed that decisions be taken by a Commission, rather than by the DG alone, and that an appeal system be established. We believe there is merit in both Oftel's and BT's proposals. We support Oftel's proposals for a regime of prohibition of anti-competitive behaviour. And we also find BT's argument for a Commission persuasive. But appeals against the Commission's decisions present more difficulties. *Prima facie*, natural justice would seem to demand rights of appeal. But appeals would delay the implementation of binding judgments and in this domain, as in others, justice delayed is justice denied. Moreover appeals are costly. Accordingly, we propose that there should be a right to appeal only in limited circumstances. John Kay's (1995) proposals are persuasive. Kay argues that appeal should be permitted only when the regulator has handed down:

- bizarre decisions
- a series of consecutive decisions biased in the same direction
- decisions on issues which are central to the regulated sector.

Kay has also argued persuasively that competition policy should focus on cases of harm to consumers or competitors rather than on whether or not a firm has a dominant position. Defining dominance is difficult and particularly so in new and converging markets. However, focusing regulatory attention on the harm caused to *the public* by restrictive practices points regulation in the right direction: that is, towards protecting all stakeholders in communications. UK regulation has been dominated by the economic concerns of the players rather than the welfare of consumers and citizens.

Quality indicators

Publication of fuller information on quality of service indicators and on usage, costs, prices and revenues for the whole telecommunication system is required if the public interest is to be served and regulation is to be effective. We recognize that much information is commercially sensitive and that collecting and disseminating it is not cost-free. Maintaining the 'any to any' principle in the context of asymmetrical competition and securing a smooth and equitable transition from plain old telephone service to a cornucopia of services delivered on the information superhighway require more information for the regulator and the consumer.

We echo Mitchell and Milne's (1990) demand for quality of service indicators to be published by all licensed public telecommunication operators at appropriate intervals (and not less than once a year). Information should be published for:

- service provision
- fault incidence
- fault clearance
- call connection
- public call-boxes
- operator services
- voice and data transmission
- billing accuracy.

The improvement in BT's responsiveness to customers since privatization deserves acknowledgement. Its response to customer complaints about billing anomalies and disputes is generally good and BT now recognizes its responsibility to compensate subscribers for interruption of service. However, BT is not the only service provider. Accordingly, it is now time for generally agreed principles with which customers can seek redress for loss suffered as a result of poor service, misrepresentation or other loss caused by network and service operators. These principles should be codified and a consumer response division established within the relevant regulatory agency(ies) to provide 'one stop shopping' for consumer enquiries and redress. Without a clearly identified destination for complaints and effective regulatory jurisdiction, the consumer interest is likely to go by default and network operators may be unjustly held responsible for deficiencies which are not of their making. We welcome Oftel's efforts in this direction, with its publication of comparable performance indicators of the main operators in the business and residential market.

Market forces in spectrum

Central to any communication policy is the question of allocation and control of two related resources: the radio frequency spectrum and the orbital positions in which communication satellites can be 'parked'. The global orbit/spectrum resource is allocated to governments by international agreement through the International Telecommunications Union, a United Nations agency. Governments then allocate their share of the global orbit/spectrum resource to national users according to national priorities. In the UK the Radiocommunications Agency (RA), now an executive agency of the Department for Industry and formerly a division of the Department for Industry, allocates spectrum to UK users. Both nationally and internationally, the orbit/spectrum resource has largely been allocated unpriced on a 'first come, first served' basis.

The orbit/spectrum resource is finite, but technological change continues to expand the limits of the resource allowing both a more efficient use of current spectrum and the use of new spectrum frequencies. The radio frequency spectrum is easily polluted: if users do not have an exclusive right to use a particular portion of the spectrum in relevant location(s) then interference between rival transmissions using the same frequency is likely to pollute the spectrum and make it useless. It is also non-depletable: when one user ceases to use the spectrum the resource remains for other users. On the other hand the orbit resource is depletable: when a satellite occupies a particular orbital position that location is not available for another satellite. Moreover, satellites have a finite life and 'dead' satellites occupy scarce orbital parking slots which cannot be used by new satellites unless the original satellites are moved.

Efficient use of the orbit/spectrum resource confers powerful social and economic benefits. It makes point to point, or point to multipoint, communication over short or long distances potentially cheap, easy and reliable. Increasing demands on the orbit/spectrum resource have stimulated technological development, with striking recent growth in the usable spectrum through the effective commissioning and implementation of technologies for multiplexing, digitalization, communication via satellites, and use of higher and higher frequencies for communication. They have also made allocation of the orbit/spectrum resource a key issue for governments.

Different spectrum bands have different uses (see figure 1). The relatively low frequency band between 28 and 470 MHz was one of the first sections of the spectrum to be exploited effectively for communications and is one of the most intensively used parts of the spectrum. In the UK more than half the radio spectrum (54.1 per cent) in this band is allocated to the Ministry of Defence (MoD). The next biggest single user is the Civil Aviation Authority (CAA) (6.2 per cent), followed by the Home Office and Scottish Office (i.e. emergency services such as police, fire, ambulance and coastguard) who account for 6.1 per cent. Land mobile communications occupy 13.9 per cent of the band, the BBC 2.1 per cent, independent local radio (ILR) a further 2.1 per cent and so on (Radiocommunications Agency 1993 p. 12).

Other bands show a different profile of users but similar patterns:[8] the CAA and media and communications firms are prominent users. In the 9 kHz to 1 GHz band, for example, defence accounts for 28.8

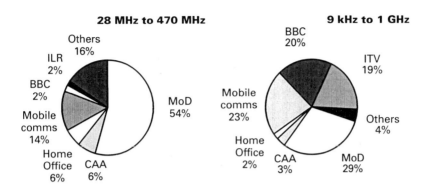

Figure 1

per cent, the BBC for 19.9 per cent and ITV for 18.8 per cent (Radiocommunications Agency 1994).

Latterly, the historical pattern of national and international allocations has been criticized from two perspectives. In an international context, 'first come, first served' allocation of the resource tends to advantage wealthy and technologically developed users whose occupation of desirable parts of the orbit/spectrum resource reinforces their advantages. Other potential users tend to be forever excluded from access to the resource. Second, critics charge that in both national and international contexts, the unpriced allocation of a scarce resource means that users of spectrum and orbits have no incentive to use the resource efficiently and thus a scarce and valuable resource is wasted.

Establishing spectrum markets

A possible solution to these problems is to create property rights and establish a market in the orbit/spectrum resource. With tradable rights to orbit/spectrum, all countries might be allocated a share in the resource, avoiding the unfairness of the 'first come, first served' principle. If they received a share surplus to their own requirements, the surplus could be traded. Clearly, implementation of such a novel system would not be easy. How should the historical patterns of allocation and usage be unravelled? On what principles should the orbit/spectrum resource be allocated? Should all states receive an equal share? Should allocation be in proportion to population, or geographical area, or something else?

The establishment of property rights and a market in the orbit/spectrum resource has been advocated vigorously in various national contexts for its potential to reduce barriers to entry to communication markets and promote efficient use of spectrum. Potential new entrants are now unable to secure spectrum allocated to users with long established 'squatters' rights' and reduce waste of resources by promoting efficient use of the orbit/spectrum resource.

Two kinds of objection have been made to proposals for spectrum markets. First, spectrum pricing will exclude deserving users who are unable to pay market rates, and socially valuable uses of spectrum, such as emergency services, defence, or community and public service broadcasting, will be 'crowded out'. And second, the costs of establishing and enforcing property rights in spectrum and in trading rights will exceed the benefits conferred by markets in the orbit/spectrum resource (see, *inter alia*, Melody 1980). Recent US auctions, however, seem to weaken this second argument: in 1994 and 1995 the US Treasury has gained a total of $9 billion (£7.2 billion) from sales of spectrum (*Financial Times*, 15 March 1995). An additional problem is how to price the spectrum for auction before its destination is decided, as value depends on the use to which the spectrum is put. The minimum/maximum price varies if the spectrum is destined for public uses (such as fire, police, defence etc.) or highly remunerative telecommunications and broadcasting applications. The problem is particularly relevant for digital broadcasting transmission which allows dynamic use of the spectrum via multiplexing techniques.

None the less, if a market in spectrum has manifest disadvantages, so does allocation through administrative decisions. Prevailing spectrum allocations reflect past priorities and the historical development of the use of radio frequencies rather than current needs. Moreover, granting spectrum to users at zero cost gives incentives to hoard spectrum rather than to use effectively an increasingly valuable resource. Interestingly, there are signs of a market system developing spontaneously in the UK public sector. West Yorkshire Police has sold to Cellnet and Vodafone spectrum allocated to it but which exceeds its own requirements. West Yorkshire Police's resale of spectrum is clearly preferable to the spectrum being unused, and thus wasted, but it is not clear why an entrepreneurial police authority, rather than the Treasury, should receive the benefits of spectrum resale.

There is no easy solution to the problems highlighted above. Indeed, the move towards a market in spectrum is likely to be partial and gradual, bringing larger and larger parts of the spectrum into the market. When analogue broadcasting is switched off, for example, large amounts of the frequencies presently used by terrestrial broadcasters will be available for other uses. Some portions of the spectrum will always be reserved to special uses, but the allocation of reserved spectrum is not incompatible with the freedom to trade any unused frequency for short lengths of time. Use-specific auctions and/or shortening of the licence period could be used to improve assessments of the commercial value of the frequencies for sale. They could also help establish whether the public users who access spectrum for free or at low price are systematically overbidding and reselling. If this is the case, their allocated portion should diminish at the subsequent round. Spectrum sale should be effected through the auction of licences to use or trade large blocks of spectrum for a finite time. Purchasers of spectrum licences, who are likely to be large and well capitalized organizations, would then sub-license portions of the spectrum thus acquired to smaller users. We might imagine the BBC and British Telecom acquiring spectrum on their own account, and possibly sub-licensing portions of their licensed capacity to third parties, and spectrum retailers acquiring licences to portions of spectrum exclusively for resale to third parties. This mechanism would promote efficient usage of a scarce resource and increase the financial return to the Treasury, and thus to the public, from an asset which has hitherto been under-exploited and whose benefits have been disproportionately enjoyed by private rather than public interests. It would lower barriers to entry to communication markets and reduce the waste of a scarce and valuable resource.

A complete transition to allocation through price may lead to the curtailment, or even cessation, of valued services because of the costs of access to spectrum. Such undesirable outcomes are most likely for applications requiring large quantities of the radio frequency spectrum, such as television, which is notably spectrum hungry. Indeed any changes to spectrum allocation and use which affect broadcasting are likely to present particularly sensitive policy problems. If, for example, broadcasters are priced out of access to the frequencies now used for broadcasting they may adopt different methods of delivering services and/or different frequencies and transmission protocols (e.g. transmitting digital rather than analogue signals) which may require viewers and listeners to acquire new receiving apparatus. Under these circumstances the costs of reallocating frequencies will thus be borne, in part or whole, by consumers who, although significant stakeholders, have no means of influencing the allocation of spectrum via a market. The imminent transition from analogue to digital broadcasting will pose many dilemmas for policy makers; we discuss this issue in chapter 2.

Not all potential uses of spectrum are equally socially valuable and government has to ensure that desirable uses are not crowded out from the spectrum. Access to spectrum for socially desirable users should continue either through direct allocation of spectrum, on the lines now used for police, ambulance, fire services and the like, or by endowing users whose access to spectrum is socially desirable with the resources to buy spectrum. Initially, direct allocation might be the simplest option to follow. This might be done by giving existing users rights to the spectrum they have historically enjoyed. However, the important change is that they would be free to continue to use this spectrum or to sell it. Some institutions would clearly enjoy a gratuitous 'windfall', but the public would not be disadvantaged, because the 'windfall' would derive from the more efficient exploitation of a resource the waste of which was hitherto invisible. Such a system would also give existing users of spectrum incentives to use spectrum more efficiently and/or cede the use of under-used spectrum to other users. New public sector users, who for reasons of public policy were deemed to be essential users, would receive resources to buy spectrum in the same way that essential services receive resources to buy essential items. The military, for example, have a finite budget to buy vehicles and therefore have incentives to buy and use vehicles efficiently. The same arguments should apply to spectrum.

There is another small category of spectrum users who are neither profit distributing private companies nor public sector organizations. These small groups of private users, such as radio hams (licensed amateur radio operators), should not be disadvantaged by changes to the system used to allocate spectrum – particularly because to do so would be inimical to the principle of opening access for ordinary citizens to the means to communicate. Their interests could be secured either by maintaining the area of spectrum designated for amateur use under international agreement outside the market system of allocation or by establishing a purchasing agency, funded by the exchequer, to bid for spectrum and allocate it to appropriately designated users.

Summary

1 Competing telecommunications networks and service providers in a single market are novel. The USA and the UK pioneered new ways of managing the telecom industry in the 1980s. Before that, monopoly provision was the norm.
2 Liberalization of the telecom markets showed remarkable results: prices went down, the quality and range of services on offer increased. However, this has been achieved at the cost of increasing regulation by an independent public body. Managing national telecom has shifted from running a centralized hierarchical network to managing a complex web of interconnected networks.

3   There are two models of telecom competition. The US established 'common carrier' competition, whereby operators use the public network to offer competing services. The UK promoted instead competing infrastructures. We believe the benefits of infrastructure competition are superior to those of a common carrier system: notably greater threat to the incumbent, incentives to invest in high return sector, greater innovation and dynamic relations with manufacturers of telecom equipment.

4   In the UK, the regulator Oftel has some powers to curb anti-competitive behaviour. We believe they are not enough. An EU-style prohibitory competition regime should be introduced in the UK, as well as wider powers for Oftel. This should be balanced by greater accountability for the regulator and a system for appeal.

5   We support the introduction of performance indicators for competition operators, which should be published at regular intervals. Experience has shown that insisting on comparable figures from the early stages of the market saves significant amounts of time and effort.

6   Some countries have also introduced market forces to the allocation of spectrum, but not the UK. We argue for some portions of the spectrum to be offered for sale. Spectrum rights for reserved frequencies should be made tradable and reallocated periodically, in order to monitor use of this public scarce resource and offer incentives to reduce wastage.

2 Essential Facilities, Third Party Access and the Problems of Interconnection

Convergence of technologies is slowly bringing distinct markets to converge. Already a number of new services, like electronic newspapers, cross the traditional boundaries between sectors. Such converging markets present new problems for regulators and confront operators with new issues. Increasingly, the ability to control a number of adjacent markets will be dependent on a few gateways. The most obvious of these is the telecommunications network. Publishers, for example, will have to face the issues of access to telecommunications network, if newspapers are to go on-line. Other bottlenecks are less obvious: the software that allows customers to choose from large screen based menus (navigation systems), the systems that authorize unscrambling of TV pictures when payment has been received (subscription management systems) or even a new system for safe payments on the Internet.

This chapter analyses some of the most controversial issues in broadcasting and telecommunications. We believe that the market developments already visible point to an even greater relevance of these same issues in the future. In this area more than any other, the ability to apply a coherent vision of the future of communications across all its segments can make a real difference. Co-ordinating the action of different authorities and government agencies is a necessity, if regulation is to modify the behaviour of a few global players. Without such co-ordination, powerful firms can exploit regulators' lack of communication and information to the detriment of competitors and the general public.

Regulating bottlenecks: the essential facilities doctrine

Technological and organizational change poses great problems for competition authorities. Defining the 'relevant market' – a crucial analytical task when considering whether a firm has abused a dominant position – is very difficult when old markets converge and new ones emerge. Moreover, the characteristic regulatory response of imposing structural constraints on dominant firms is often at odds with the need to allow firms to find their own shape during phases of transformation. For example, regulation prevented the railways in the United States from adapting to the shift to road transportation; neither the public's nor the railways' interests were so served. It is preferable to regulate behaviour rather than structure particularly when the regulated sector is dynamic. Structural regulation seeks to act upon the *number and size* of the players in the relevant market. Behavioural regulation focuses on prevention, and punishment, of anti-competitive *behaviour* whatever the size of the regulated firm. Behavioural regulation is better adapted than structural regulation to dealing with a class of problem that is likely to become increasingly important in media and communications: the control of bottlenecks and gateways. A firm may not be dominant in a given market but may still control third party access to facilities essential for other firms to compete in the same or a contiguous market. Interconnection is one important instance of the general problem of regulating third party access to essential facilities but is far from the only relevant instance. Increasingly, the focus of regulatory concerns will be on access to:

- conditional access systems
- subscription management systems
- navigating systems
- new and intelligent broadband networks.

If these problems are likely to become more general, rules are required to deal with the *class* of problems rather than rely on *ad hoc*, case by case, decision making – a procedure which is unlikely to provide either prompt decisions or guidance to firms contemplating the introduction of new services. The essential facility doctrine, pioneered in United States regulation and increasingly adopted by the European Union, provides a basis for rules to regulate third party access to facilities. For the essential facilities doctrine to apply, these conditions must be satisfied:

- The facility must be an infrastructure or a combination of infrastructure and service.

- It must be practically or reasonably impossible for competitors to duplicate the facility.
- It must be feasible to provide the facility to competitors.

If these conditions are satisfied and if denying access to the facility would make 'competitors' activity in the market in question either impossible or permanently, seriously and unavoidably uneconomical' (Temple Lang 1994 p. 488) then a firm may be required to provide third party access to the facilities in question on fair terms. There are obvious second order problems involved in the application of the essential facilities doctrine (notably in determining what are fair terms) but they are, we believe, insufficient grounds for rejecting the essential facilities doctrine as a basis for settling conflicts over third party access.

The European Commission has resolved a number of analogous cases by using the notion of essential facilities. For example:

- In the operation of port services, the owner and operator of a port was required to grant access to its competitors on terms no less favourable than those which it gave to its own ferry services. Terms were defined in terms of price and the position and relative attractiveness of different berths within the port. This judgment provides a useful basis for arbitration in cases where broadcasters dispute their locations in browsing system 'front ends' which are likely to be required to make accessible and usable the hundreds of channels promised in digital satellite broadcasting (see *B&I Line* v. *Sealink Harbours Limited* [1992] 5 CMLR 255).
- In the air transport market airlines have been required to grant interlining facilities to other airlines. This judgment provides a useful basis for arbitration in cases where interconnection of facilities is disputed (see *British Midland Airways Limited* v. *Aer Lingus PLC* [1993] 4 CMLR 596).

More effective regulation to deter and penalize abuse of a dominant position and to ensure access to essential facilities will do much to ensure the development of new media and communication services and to protect the interests of consumers. But stronger competition laws and regulation to facilitate third party access to essential facilities, though necessary, will not be sufficient to resolve all problems.

Conditional access

Conditional access is especially important because conditional access systems (CAS) are going to be at the heart of new media such as video-on-demand, on-line services and digital television. A conditional access system is 'an effective and secure subscriber authorisation mechanism so that only those viewers who have paid for a programme or service are able to receive it, i.e. there is only conditional access to the programmes' (ITC 1995a). CASs include both encryption (scrambling of signal and the key to unscramble it) and subscriber management systems which establish viewers' entitlement to receive the programme and which are used to collect subscriptions. Conditional access systems are costly, both for the developer and for the consumer who buys the set-top box connecting the CAS to his or her television set. Although any broadcaster could produce its own CAS, there are clear benefits if consumers are not segmented into rival systems and need buy only one box. In consequence, first mover advantages in these markets are tremendous, and a firm which controls conditional access is likely to have gatekeeping powers over the entry of new services to the market. It will also have the power to set prices and conditions and, potentially, to abuse that power. Control of conditional access has implications, therefore, for pluralism and competition, for consumer prices and for a host of other issues affecting the public interest. In the past conditional access systems might have been seen as an obvious candidate for public ownership. But no contemporary government is likely to commit the estimated ECU500 million required to build a CAS infrastructure and choose among competing new technologies.

The European Union has backed the Digital Video Broadcasting (DVB) group, an association of 150 interested operators, manufacturers, broadcasters and regulators, hoping to avoid industry battles and fragmentation of markets. The DVB group gathered to 'create in Europe a framework for a harmonious and market driven development of digital television by cable, satellite and terrestrial broadcasting' (House of Commons Library 1994 p. 31). The group, however, failed to establish a European CAS standard. Advocates of a unique proprietary standard (Simulcrypt) and proponents of a mandated open standard (Multicrypt) failed to agree. Adoption of Simulcrypt, backed by the existing subscription television providers (notably BSkyB, FilmNet and Canal Plus), would offer important advantages: there would be a single set-top box and a single smart card. The dominance of a single proprietary standard would simplify access to customers at the cost of substantial market power ceded to Simulcrypt firms. By contrast Multicrypt, backed by a number of broadcasters including the BBC, would allow different CASs to be used with the same set-top box, where different smart cards unlock different systems. Adoption of the Multicrypt system would favour competition, but would necessitate substantial investment by its sponsors with an uncertain prospect of financial returns. Who would spend the

ECU500 million required to commission Multicrypt without being able
to capture proprietary revenues from its deployment? Would manufacturers and consumers prefer Simulcrypt or Multicrypt? Consumers have
shown a fear of conflicting proprietary standards following mistaken
purchasing decisions in earlier 'standards wars' such as those between
Betamax and VHS, Apple and IBM, Sega and Nintendo etc. However,
although consumers' attitudes may seem to favour Multicrypt,
Multicrypt would mean that consumers would have to change smart
cards when navigating between subscription services using different
CASs and would be billed by more than one supplier. Given that there is
no clear balance of advantages between the systems, consumers may be
best placed to choose. The advantages and disadvantages associated with
each standard are summarized in table 2.

As the DVB group failed to establish a consensus, the European Union
devolved choice between CASs to the manufacturers. The Commission
of the European Communities however did address broadcasters' concerns by amending the EU Directive on Television Transmission
Standards (DIR 95/47/EC) and requiring that:

- Conditional access providers should offer services on a 'fair, reasonable and non-discriminatory basis'.
- Providers should present separate accounts for broadcasting and conditional access activities.
- Manufacturers should not be prevented, deterred or discouraged from
  including a common interface in equipment.
- There should be an easy, speedy and inexpensive dispute resolution
  mechanism.

The failure to agree to just one system is likely to be only partially offset by these amendments. In particular, the ability to resolve disputes
speedily will prove to be the crucial point. General competition law, as
well as the essential facility doctrine, can be used to correct abuses and
rectify situations in which third party access has been denied. Although
it will prevent abuse of market power, competition law cannot be used to
create a level playing field. Policies aimed at ensuring that all firms enter
a new market enjoying the same advantages cannot peg on anti-trust provisions. They must be an explicit policy goal of the regulator. Should the
UK regulator judge digital television to be a market where an equal start
for all firms is required, then different tools are needed. Rate of return
regulation of the business providing conditional access would be the
most appropriate type of control in this case, as it avoids the problem of
setting a 'fair price' for the conditional access service.

Table 2

	Cost	Piracy	Advantages
Simulcrypt	At volume, a set-top box will cost £200 and the smart cards about £5	Better protection against piracy: smart cards would always be inserted, and licensing can be better monitored with closed standard	Better fit to consumer needs, hence facilitating market for all broadcasters
Multicrypt	The smart card would cost £20–30, but it would be much cheaper to buy a new card than a new set-top box	Cards, rather than set-top boxes, need replacing if pirated	Lower barriers to entry into conditional access systems

Spectrum and the transition to digital broadcasting

In chapter 1 we have argued for some portions of the frequency spectrum to be subject to market forces. In broadcasting, the difficulties of effecting a market driven transition are compounded by the need to balance the right of spectrum owners to sell their frequencies with the interests of current producers and consumers. In particular, broadcasting's current universal reach needs to be maintained. Digitalization further complicates the problem because as long as analogue and digital broadcasting co-exist, spectrum requirements will vary widely for different broadcasters. The current allocation of frequencies is based on a bargain between government and broadcasters defined in the terms of the licences granted to broadcasters. The government allocates spectrum for a specific end-use (broadcasting) in return for the broadcaster's commitment to obey certain regulations. Broadcasters must observe the codes of practice and taste and decency rules defined by the regulators (the ITC, Radio Authority). Terrestrial advertising financed television services (Channels 3, 4 and 5) have positive programming requirements imposed on them as conditions of licence. The BBC's Charter, and Licence and Agreement, define its responsibilities. These public service content requirements, together with the universal availability of terrestrial radio and television signals, constitute the 'community service obligation' implicit in the present broadcasting order. However, this bargain was set up when terrestrial television was the only delivery system, and it can work as long as terrestrial delivery is preferable to cable or satellite, in terms either of reach or of costs. Consequently, the nature of this bargain has changed, and will continue to change, as alternative methods of distributing radio and television services develop.

Until the 1980s, broadcasting services in the UK were delivered, with some minor exceptions, by analogue terrestrial transmission. Satellite and cable distribution of analogue signals started to become significant in the 1980s and grew in importance in the 1990s. About a fifth of British households have cable and satellite services in the mid-1990s. Although terrestrial reach is expected to maintain a significant lead over alternative systems in the next decade, comparisons between various systems are not straightforward. Terrestrial broadcast (analogue and digital) is the only delivery system which can provide portable reception. Cable is the only system which can provide full interactivity. Satellite represents the cheapest way to bring television signals to remote areas. Also, a broadcast TV signal needs 2,000 times the bandwidth that a telephone call requires. Since spectrum is scarce but the capacity of a fibre optic cable system is virtually unlimited, some believe that spectrum will eventually be used only for those applications which need delivering without wires (for example mobile telephony). This argument, sometimes dubbed the 'Negroponte switch', predicts that future television may be wired and telephones wireless. In the words of the author: 'the bandwidth available in the ether is scarce by comparison with that provided by fibre and our endless ability to manufacture and lay more and more of it. I propose a trading of places between the wired and wireless information of today' (Negroponte 1995 p. 24).

Given the uncertainty about the pace and extent of the Negroponte switch, governments should be extremely wary of picking television's future distribution system. This is a choice best left to interaction between producers and consumers, who will be able to balance the costs and the pros and cons of contending technologies. When allocating licences, the government has to decide to what use they will be put and for how long. Only a market in spectrum will allow the government to avoid this risky decision. The uncertainty of technological developments and take-up rates suggests that a market driven transition will ensure that spectrum is put to the most efficient possible use.

However, a change to a market in spectrum could have profound consequences. Formerly the government was able to exert considerable power over commercial broadcasters by setting the terms on which access to spectrum was granted. The growth of alternative distribution systems and the greater availability of spectrum – which digitalization promises – imply that government's power will continue to diminish. Broadcasters required to satisfy positive programme requirements in exchange for access to terrestrial transmission frequencies will be disadvantaged in comparison to broadcasters using alternative distribution media, like cable and satellite. They may prefer using the other systems.

Realizing the public interest in broadcasting requires compromise between a market in spectrum, with its benefits of efficiency and openness to new entrants, and the need for the government to use its control of spectrum to secure the community service obligation in broadcasting (see chapter 4 for discussion of this concept). The public interest will best be served by reserving a section of spectrum for delivery of community service broadcasting. Furthermore, there is a general public interest in a speedy transition to digital transmission which government can play an effective part in accelerating.

Switching from analogue to digital transmission offers numerous potential advantages. First, digitalization will use spectrum more efficiently and thus permit more services to be provided, whether broadcast television and radio or mobile telephony. Depending on the type of information transmitted, digitalization will permit between two and eight times as many programmes to be transmitted digitally using the same spectrum capacity as a single analogue channel. However, the benefits of digital broadcasting vary with different types of delivery.

The promises of digital broadcasting implied by the different delivery systems are compared in table 3. The main differences concern the number of channels potentially available and the portability of the transmitting device (aerial, dish or cable) to all TV sets in the household. As the table points out, cable has potentially unlimited capacity, and the only bottleneck to its expansion is the original architecture of the system, which will need adapting as capacity increases. The important implication is that a cable system could relay all digital terrestrial services, plus cable-exclusive services, without running short of capacity. The potential reach of cable, however, is not unlimited: unless explicit obligations are set in place, market forces would only wire up to 85 per cent of the country.

With all transmission systems, digital broadcasting will improve image and sound quality and permit novel and sophisticated services, such as traffic reports and emergency information addressed to specific users. For consumers, the advantages of digital broadcasting are:

- better reception of free-to-air TV, which will be robust even with set-top antennas
- wide-screen images with the appropriate TV set
- much greater choice of channels and services with all transmission systems.

The main disadvantage is the need to change the television set or buy a digital set-top box. The pioneering consumers will pay a premium for their early take-up: the price of the set-top box will inevitably be higher until mass market production is reached. Estimates price the early set-top boxes at £500, dropping to £200 at mass production stage. The first batch of digital TV sets will cost around £1,000.

For all broadcasters, digital technology has the key advantage of allowing an explosion of supply. However, new and costly transmission systems are needed. Also, free-to-air television will see much fiercer competition, as many more channels battle for viewers' attention and advertising revenues. This could threaten the viability of the advertising funded elements of the present public service broadcasting regime and might lead to a reduction in the number of services available free at the point of use and/or the quality of the services they provide. The BBC, so long as it continues to improve its efficiency of operation and to receive adequate licence fee funding, is unlikely to be seriously threatened. The strongest challenge will be to the advertising financed television Channels 3, 4 and 5.

Table 3

	Digital terrestrial simulcast	Digital terrestrial (after switch-off)	Digital cable (fibre optic)	Digital satellite
Number of channels	18–20	Up to 368	Potentially unlimited	150
Scarcity limit	Spectrum allocated	Spectrum allocated	System architecture	Orbital position
Maximum reach without explicit USO	(analogue 99.4%) Dependent on bids	Dependent on bids	80–5%	70%
Consumer drawbacks	New set-top box	New set-top box, perhaps new aerial	(New set-top box) Not suitable for portable televisions	New set-top box and aerial attachment Not suitable for portable televisions No regional variations

The ITV has suffered more than the BBC or Channel 4 from the growth of satellite and cable television. It remains the medium of choice for many advertisers and continues to enjoy buoyant revenues even though its share of viewing has declined. However, as satellite and cable viewing increases, the relative attractions of Channel 3 (and Channels 4 and 5) as an advertising medium will decline. Then the bargain which supports delivery of community service broadcasting and ensures delivery of positive programming requirement by commercial broadcasters will become less attractive to broadcasters. Why continue to provide expensive programming and economically sub-optimal scheduling if the countervailing advantages conferred by privileged access to terrestrial broadcasting frequencies become less significant? If terrestrial distribution is to remain the medium of choice, there must be a rapid rollout of a pervasive national digital transmission system. This means that the heavy costs of building the system must either be assumed by the government or guaranteed by it. Embedded within the Conservative government's proposals for the BBC and for digitalization there is a messy policy for doing just this. Receipts from the sale of the BBC's engineering and transmission divisions are earmarked to defray the costs of the BBC establishing national digital services.

The government has proposed to maintain analogue services alongside the new digital ones, until 50 per cent of homes are equipped to receive digital services or after five years, whichever is sooner (Broadcasting Bill 1995). It will initially allocate only portions of spectrum which cannot be used by analogue systems, the so-called 'taboo frequencies', for free. Frequencies will be bundled up in units, the multiplexes, which can be used fully or divided further and traded or shared. The spectrum capacity required for a particular signal will vary from time to time; for example, a static studio talk show will require less bandwidth than one showing rapid movement like a sports programme. Multiplexing will permit more signals to be carried and spectrum to be used more efficiently. The government proposes to grant an entire multiplex to the BBC while the other terrestrial broadcasters will be granted half a multiplex each, with Channel 4 and S4C sharing their half.

We endorse the government's commitment to ensure that there is a period of 'dual illumination' where digital and analogue services are both provided. Balancing the rival goals of ensuring efficient use of the spectrum, served by a speedy termination of analogue services, and ensuring that viewers and listeners can amortize the costs of their analogue equipment over a long period, will not be easy. However, we anticipate that pressure from potential spectrum users is likely to be more focused and vociferous than will be the pressure from consumers for continued provision of analogue services. Here, there is evidence of the need for a body to focus and advocate consumer interests in media and communication policy.

The problems with the government proposals begin when conversion is analysed from the consumers' perspective. Here, the real policy dilemma is in ensuring that consumers are not required to buy more than one set-top box to enjoy both free-to-air and subscription services which come with different delivery systems. Existing digital standards are different for each delivery system: there is no guarantee that manufacturers will supply digital set-top boxes which can convert all types of signal. Either the government remains neutral about this issue, and allows the market to come up with a solution, or it intervenes by choosing a technology which will lead the transition to digital. The government has done neither and both. Granting the best multiplex to the BBC, and giving it the means to pay for replacement of the analogue transmission system, are tantamount to selecting digital terrestrial as the leading technology. But the question of the set-top boxes has been left to the market. The market will have to deliver a common interface for decoding – which is likely – and a common system for conditional access – which is unlikely – in order to avoid the need for several expensive set-top boxes. Conditional access is going to become the single most contentious issue in the transition to digital TV.

The DVB group has agreed a common European standard for the transmission of digital satellite, cable and terrestrial television pictures, a common scrambling system for those pictures, and a code of conduct to govern relations between providers and broadcasters, particularly where the providers are also broadcasters. However, the group has not agreed on a common standard for conditional access systems (CASs). The lack of compatibility between different conditional access systems is not an obstacle to the delivery of free-to-air television. Even in a future scenario where the reach of cable and satellite rivals that of terrestrial, free-to-air channels are likely to be relayed by the cable and satellite operators, or 'must carry' requirements can ensure that this happens. But terrestrial subscription and pay per view services do need a conditional access system. They would be unavailable to cable and satellite households who do not invest in a second set-top box.

So far, the government has taken no position on this issue of a single CAS standard. But it has enabled the BBC to proceed to digitalize its transmission system and thus introduce terrestrial digital services. Once comprehensive digitalization has been established by one broadcaster then others can 'piggy back' on the infrastructure. But most terrestrial broadcasters plan to use the spectrum freed by digitalization to introduce pay per view services. So, the obdurate question of whether CASs should be harmonized cannot be ignored. A single proprietary CAS is a 'bottleneck' facility potentially offering its owner the power to control the terms on which services – perhaps services competing with the ones offered by the owner of the CAS – reach viewers. Yet there is a strong consumer interest in a single CAS standard, so that consumers do not risk being the victims of industry battles and needing several set-top boxes to enjoy the full range of digital services.

BSkyB, the market leader in CASs, has declared willingness to co-operate in developing a dual conditional access system (terrestrial and satellite) if allowed to maintain proprietary rights over it. This would give BSkyB an enormous amount of market power. But the alternatives are also unattractive. If government mandates a standard the standard chosen may not be supported by the industry (and in UK satellite BSkyB is the industry!). If rival standards develop, consumers are likely to lose. The government must compare the benefits of easier and greater choice for the consumer stemming from a single CAS standard, with the risk of entrusting control of a key facility to a powerful private firm.[1] Not an easy choice! Whatever the future developments, the regulation of gate-keepers will play a central role in future public policy.

The government's proposal to establish several multiplexers is also flawed. There is a powerful efficiency case for multiplexing, but it is unnecessary for there to be more than one multiplexer. Allocation of guaranteed capacity to established broadcasters will not optimize efficient use of spectrum and privileges existing broadcasters relative to possible new entrants. Accordingly, we believe it preferable for spectrum to be controlled by a central state agency (we propose that this should be Ofcom, a united regulatory agency – see chapter 8), which will reserve a portion of it for community service broadcasting. Rival potential franchisees will periodically compete for licences to provide services which satisfy Ofcom's community service requirements and which will be free at the point of use. The regime we believe necessary to secure the public interest in the digital era requires co-ordinated action by the Independent Television Commission, the Radiocommunications Agency, Oftel and the BBC. The challenges of managing the transition from analogue to digital make an eloquent case for regulatory reform and for creation of a single regulatory agency for media and communications.

It is appropriate to comment here on the controversial requirement for balancing payments to be made either to or by Channel 4. We share the ITC's view that future advertising revenues and patterns of expenditure are uncertain and that therefore it is appropriate to maintain arrangements where Channel 4, a non-profit-distributing public service channel, is guaranteed subventions from the advertising revenues of Channels 3 and 5 if, at some future time, they are required. When, as now, Channel 4 is enjoying conspicuous success in gaining viewers attractive to advertisers, subventions should flow in the other direction.

Interconnection

Setting the terms for interconnection, the linking of separate telecommunications networks, has proved to be the most difficult issue in telecoms regulation. Interconnection is the most important instance of third party access problems in contemporary media and telecommunications. Regulation is likely to be needed in the UK for a long time to ensure that the terms of access to the dominant network are fair and likely to promote competition. Rather than disappearing, regulation should gradually shift away from setting the final prices and focus specifically on the areas where competition is slowest to develop. A successful solution to the difficult issues of interconnection is a necessary condition for competition to be established at all levels. Fair interconnection terms are essential if the dominant operator's advantages inherited from the past are to be reconciled with effective competition. This requires changes because the system of interconnection, and interconnection pricing, set up during duopoly, is not appropriate for a market with many players. A new system is needed to reflect changes in the structure of the UK telecommunications market.

Until recently, interconnection agreements were based on the 1985 agreement between BT and Mercury, which set interconnection charges per minute of usage on the basis of fully allocated costs. In 1993, Oftel consulted extensively and reviewed interconnection arrangements. In early 1995 Oftel decided that the BT-Mercury agreement was to be replaced by a system whereby interconnection charges were set by reference to a published list of standard interconnection fees for each service available on the network. This short term settlement, which should be in place until approximately mid-1997, has the benefits of:

- greater clarity for a new player, compared with the previous arrangement where agreements were to be reached on a case by case basis
- no delays
- updated list of services
- contribution of all operators to its determination.

This transitional measure is intended to pave the way for a long term redesign of the interconnection agreement involving review of all interconnection issues including:

- who should be allowed to interconnect and who should be required to offer interconnection
- services for which operators can/should interconnect
- the point of interconnection

- the charging methodology
- regulation of charges
- monitoring of interconnections
- timing of the new arrangements.

The terms on which BT and other public telecommunication operators (PTOs) are required to interconnect with other operators will have a significant impact on investment decisions. The choice of services that interconnected operators will be allowed to offer over other PTOs' networks will influence both the range of services on offer and the speed with which innovations are implemented. Most UK operators imitate each other, offering very similar services differentiated by subtle variations in tariffs. The level of interconnection charges will affect the cost structure of all operators. A debate on interconnection is really a debate on the type of competition which the UK regulators want to pursue in the future.

The rules that govern interconnection have a pervasive and powerful effect in shaping the commercial relations between players and their investment decisions. The UK now has a unique opportunity to make the most of its distinctive liberalized regime. Having stimulated investments in rival networks, having pioneered the exploitation of economies of scope with other utilities and entertainment, government and regulators should now ensure that competition stimulates the introduction of innovative services and pricing options. Service competition is hampered by high prices and by disagreements on the level of interconnection. If a new phase of development in UK telecommunications is to take place a clear and equitable interconnection regime is required. The time honoured principle of telecommunications planning, 'any to any' connection, must underpin network and regulatory planning.

A new interconnection regime should be simple and transparent. The present system is ridiculously complex: interconnection rights change with the type of licence held, interconnection responsibilities increase with market share, and general rules do not apply if a retail price for the service requested is not published. Operators must adhere to the level and price of interconnection which their licence prescribes even if both negotiating parties would like to override those rules. The present regime is opaque and inflexible. To facilitate service competition, all service providers should be able to bill customers directly and distinctions based on licence status should be broad. For example, class licensees who do not own any infrastructure are allowed to buy capacity at a discounted rate and resell it, while operators with their own infrastructure can interconnect to other networks.

New entrants and incumbents have contradictory interests. Past experience has shown that operators with unequal market power may take years to reach agreement on interconnection terms. Accordingly, regulatory action is required to monitor and countervail the anti-competitive effects of delays. There is also a clear regulatory role in ensuring that a dominant player does not impose proprietary standards and thereby gain a (temporary) monopoly through the incompatibility of existing and emergent networks. Without countervailing regulatory action a network operator may require interconnecting operators to take access services as part of a package, so that operators have to purchase the entire package even if they need only a single component. Bundling of interconnection services means that, for example, rather than selling interconnection from customer to local switch and from local switch to trunk, a dominant PTO may only sell interconnection from customer to trunk network. In consequence, interconnecting services should be *unbundled* where possible, so that the buyers are free to choose the level of service they need. Ideally, all operators should have the same pricing flexibility as the dominant player.

Interconnection charges should give accurate signals for investment decisions, by clearly *reflecting costs*. This way a new entrant can make the right judgement when evaluating whether (and where) to purchase inter-connecting services or to build its own facilities. Interconnection charges should also give the *correct entry signals*, by ensuring that only com-petitors who are more efficient than the incumbent enter the market. Interconnection charges that are set too low may allow entry to oper-ators who are less efficient than those already in the market. This would lead to waste of private resources (money could be better invested else-where) and an overall loss to society. Interconnection charges should thus be a *sustainable base for retail prices* and we support the require-ment for BT to present separate accounts for its network operating business and for its retail business. Transfer prices between BT network and BT retail should be transparent and operators behaving anti-competitively should be subject to penalty.

Setting the level and price of interconnection

Telecommunications experts and operators alike agree that interconnec-tion is the most difficult and puzzling issue yet to be solved. For there are seldom either freely negotiated examples of agreements or robust theoretical models to provide yardsticks for an adjudication when inter-connection agreements have not been concluded to the satisfaction of all parties concerned. Perfect competition theory with its Pareto efficient general equilibrium[2] does not apply to telecom markets. In such markets, where economies of scale and scope are present, marginal cost pricing will lead firms to bankruptcy, as marginal costs fall with increases in output to below the level of average costs.

The theory of contestable markets, that is markets where there are no sunk costs and agents can engage in 'hit and run' competition, offers an alternative economic model. It is a theory, however, which assumes products to be perfect substitutes and retail prices to reflect costs rather than monopolistic mark-ups (see Baumol and Sidak 1994). Many policy makers find contestable markets theory inadequate for regulating the real world. Its inadequacies for telecommunications markets with large sunk costs require no emphasis.

Eli Noam has observed (see Alleman 1995) that rational profit maximizing PTOs in unregulated telecom markets would only conclude interconnection agreements if they could expand markets without driving prices down by doing so (as is the case with international services). Noam's observation is helpful in advancing beyond contestable markets theory. His insight suggests that conflicts over the level of interconnection – that is the interconnection services – each carrier should offer – could be resolved using the concept of third party neutral transmission. Rather than appealing to the customary concept of common carrier responsibilities, Noam argued that an interconnection regime which allowed operators to choose whom to connect to their network but prohibited discrimination between customers would provide a good basis for adjudication. If a network operator wants to expand and interconnect with other networks then it is required to interconnect all services provided by the other party. This, Noam argues, will lead to beneficial price competition between operators. This is an attractive model, but setting (and regulating) interconnection charges will remain controversial.

There are two main alternative methodologies for interconnection pricing: usage per minute pricing and capacity based pricing.

Usage per minute charges

The principle underlying time based pricing methodologies is that the costs of interconnecting should be estimated as the cost of providing a particular service (access and conveyance) or its components. This is the method used in most countries, although in many different variants. Its main conceptual difficulty lies in the attribution of fixed and/or common costs to a particular service. Moreover, charges can be based on different accounting conventions (historical versus current costs) or on economic costing principles.

Contestable markets theory recommends the use of an *efficient component pricing rule* (ECPR) for interconnection charges. This rule advocates a two part pricing, with one part of the interconnection charges reflecting the operator's costs[3] in providing interconnection and the other part reflecting the revenue lost, the opportunity cost, arising from the operator providing carriage but not providing the service delivered to the final consumer. If interconnection arrangements based on the ECPR are to give the right signals for long term decisions, pricing should be based on the long run incremental cost (LRIC) of providing service. LRIC methodology ensures that only firms more efficient than the incumbent enter the market because only such firms, for which the cost of providing the final product is lower than the incumbent's, can make profits. An important qualification must be made here: the ECPR assumes that hit and run competition has eroded the incumbent's profits, so that the final price is equal to the stand alone cost, that is the cost of providing a given service by itself. The opportunity cost mark-up is not meant to support monopolistic prices that the incumbent was able to charge prior to competition. Stand alone costs and LRICs provide an upper and a lower boundary to acceptable interconnection costs.

Recognizing that interconnection pricing based on the ECPR promotes efficient entry, we need to highlight its main shortcomings:

- As long as retail tariffs are averaged, the ECPR charges can be lower than stand alone costs and lead to inefficient duplication.
- In most instances the cost estimates are provided by incumbent firms, and will thus reflect their costs and inefficiencies.
- The ECPR assumes that the products supplied by the regulated industry are perfect substitutes. This is not always the case and where products are not substitutable the opportunity cost component of the price rule should be reduced.
- When prices fall the demand for telecom services grows. Hence a service provided by the new entrant might not displace any service from the incumbent. If this is the case there is no rationale for a mark-up to reflect the opportunity cost.

An alternative method derived from the theory of contestable markets is *Ramsey pricing*. This two part pricing method seeks to set final prices which lead to economically efficient levels of demand. The mark-up on LRICs required for recovering common costs are adjusted on the level of demand elasticities. Higher mark-ups are added for services with lower elasticities. The idea is that mark-ups causing equal changes in final demand cause minimum distortions. Where products are not perfect substitutes, Ramsey pricing requires considering cross-elasticities, or the variation in final demand for a telecom service caused by the variation in price of another telecom service. Although theoretically attractive, Ramsey pricing presumes cost-free access to information which, in practice, is burdensome and difficult to obtain. Most cross-elasticity estimates are based on historical data which, in an industry of fast changing services and technology, may not be a good measure for present and future demand. Furthermore, Ramsey prices do not allow for flexibility, since LRICs are estimated on the dominant operator's costs and competitors are forced to follow its cost structure with adverse consequences for efficiency and innovation.

Capacity based pricing

Tariffing methods based on capacity are used in utilities like electricity where the costs of network building depend principally on the maximum level of capacity which has to be accommodated in the network at peak time, although, on average, the capacity used at other times is lower. The appeal of such methods lies in the economic benefits they confer by linking interconnection tariffs to network costs and in the greater flexibility they offer to interconnecting operators. In practice, however, implementing capacity charging in a multi-operator environment is to venture into new territory necessitating a series of difficult choices. The choices regard the definition of peak time adopted, the determination of the network path to be used, the charging units and so on – and in particular whether charges are to be based on the amount of peak time traffic *actually used* or on the amount of capacity *booked* by the operators. Pricing based on actual usage would ensure the highest level of economic efficiency but would also require a great deal of information collection.

The peak time in telecoms is the time when additional investment is required for more traffic to be carried by a given system. It is given by a *daily* busy hour, rather than a seasonal peak time as in the supply of energy, where peak time is defined as the 'coldest half-hour between November and February'. For usage based pricing actual two way telecom traffic would need to be estimated and checked continuously. In addition to this difficult and costly exercise, there are also uncertainties about the actual level of costs for which interconnecting operators would be liable. Operators would only know their interconnection payment after the periodical definitions of busy hour have been made. This methodology is unlikely to encourage competition and innovation in new services. If entry is to be promoted under such a pricing regime, and the risk to new entrants reduced, operators would have to make traffic data public. Operators are unlikely to welcome the requirement to disclose this commercially sensitive information! Moreover, checking levels of under- or over-booking of capacity constitutes yet another task for regulators and operators and thus another cost levied on users.

This brief overview shows the complexity of interconnection pricing decisions in telecoms. Regulators may safely conclude that interconnection charges should be no higher than stand alone costs and no lower than incremental costs, but determining these costs is practically difficult and necessitates decisions on several vigorously contested theoretical issues on which reasonable people can reasonably disagree. The main difficulties arise from these factors:

- Telecommunication products are not perfect substitutes: some services are close substitutes, some are complementary, and some are inputs to further services.
- Retail tariffs are highly averaged, to provide for loss making services and customers.
- The allocation of costs to one particular service is always controversial, as only a fraction of total costs can directly be attributed to one service or another.
- Competition itself shapes the relative importance of different cost drivers.

What is the best course of action? UK experience has shown that regulatory intervention is needed. British Telecom and Mercury were unable to reach agreement on interconnection terms during the period of duopoly. Delays in negotiations are harmful to new entrants. But the imposition of an unsatisfactory solution is not a viable regulatory recommendation. There is also powerful testimony from New Zealand, where no industry-specific regulator has been set up, that an unsatisfactory, albeit prompt, settlement is worse than a settlement delayed. New entrants concluded that regulation offered a cure worse than the disease it was created to relieve: 'Even a liberalised market which has relatively long delays and high transaction costs achieves better outcomes than regulation' (Davies 1995).

On balance we believe that neither regulation nor unregulated competition reliably produces satisfactory interconnection agreements but, none the less, we believe an industry-specific telecommunications regulator is essential for the UK – not least because the inheritance from the past has left gross asymmetries in the size and power of competing firms. To resolve the conundrums of interconnection, we believe that the best, albeit imperfect, way forward is to elaborate a framework of principles which specify who has a right to interconnect for what services, and impose a time limit within which commercial negotiations may take place. Specifically, we recommend that the regulator establishes:

- provisions to avoid delaying tactics
- default terms and conditions that can be overruled by private agreements
- a price cap on interconnection charges

These provisions are to be supplemented by a strengthened competition policy which, among other things, will ensure that interconnection charges should be at rates no higher than those operators charge their own retail operations.

Summary

1 Third party access to networks and gateways is likely to become an increasingly important issue for regulators in the future. Rather than dealing with these issues in a case by case manner, regulators' decisions should be based on the essential facility doctrine.

2 The essential facility doctrine can be applied to any firm that owns or controls a bottleneck which is necessary for competitors to operate in a given market. Although the gatekeeper is likely to have some market power, it need not be dominant. Nor is it necessary to prove that the essential facility is a natural monopoly.

3 Conditional access systems (CASs) are a particularly important case in point. Failure to agree to a single EU standard presents policy makers with the dilemma of either conferring enormous market power on one gatekeeper or condemning consumers to rival incompatible systems.

4 The CAS issue is also central to managing the transition to digital broadcasting. In the digital age, the government will still need to compensate community service broadcasters, who are subject to positive requirements, with attractive spectrum. To do so, carriage of community service broadcast must be mandatory under 'must carry' rules.

5 Interconnection is the most difficult issue in telecommunications. There is no satisfactory theory or commonly accepted practice to guide regulators. While the principle of 'any to any' interconnection is widely accepted, the details of its implementation remain controversial.

6 Usage per minute charges and capacity charging are the two existing charging methodologies. Within each type, endless variations are possible. Although capacity charging is very attractive in theory, its practical implementation may well prove impossible in a multi-operator environment.

7 We recommend that the regulator acts against delaying tactics in interconnection and offers a set of default terms and conditions when commercial agreements fail to materialize in time. Above all, the existing complex system of licences and related interconnecting rights and responsibilities should be simplified.

3 Concentration of Ownership

Impartial and accurate supply of information is fundamental to the well-being of the 'marketplace for ideas'. Yet we accept that impartiality and objectivity are ideals rather than achievable goals, and we demand therefore that what cannot be achieved by individuals should be approximated by the market as a whole. We expect the media markets to provide us with plurality of sources and diversity of content. Market forces alone cannot be trusted to deliver these two goals; the market imperfections discussed earlier (see introduction and appendix 1) point towards increasing concentration in the media sector. Some concentration of economic power can be desirable: risky new media markets require venture capital and market power in order to launch new products. And the public interest in authoritative and accurate reporting seems to be better achieved by media which dispose of significant resources. Too many small information providers may not provide the depth and width of coverage which form part of the public's media entitlements. But two chief potential harms may result from concentrated ownership and control of the media:

- Some important viewpoints may be unrepresented or under-represented (and others over-represented).
- An interest group(s), whether political, commercial or otherwise, may use its media influence and power to exercise political influence and power.

The first is harmful to the goal of diversity of content, the second to the goal of plurality of ownership. We will argue that these two objectives cannot be achieved simply by acting upon concentration in the media markets.

Market structure, and hence ownership, is an important question, but only one among several relevant policy questions. The goals of effective competition, impartial and accurate supply of information, diversity and plurality are not to be achieved simply through getting ownership right and expecting achievement of other goals to follow automatically. Not only is there a complex interrelationship between advertising and final consumption markets in the commercial media, but other issues are also significant, such as the professional values and practices of journalists; the degree of independence enjoyed by editors; and the relationships between sources of information and information media, governed by matters such as privacy and libel law, official secrets and freedom of information legislation and traditions of accountability of the executive to the fourth estate.

Our proposals on ownership cannot be separated from our consideration of the broader regulatory framework for the media, so we begin here by looking at policies for competition and democracy in general. In that light, we examine current UK regulation on cross-ownership and concentration of ownership. We focus specifically on television, newspapers and radio as these media present the clearest instances of the difficulties in navigating between the contradictions of economic opportunity, and democracy and a strong civil society.

Competition: more behaviour than structure

Fair competition between many providers is widely seen as the key to unlock both an effective marketplace for media goods and a democratic marketplace for ideas.

The goal of competition policy is to prevent anti-competitive behaviour, that is, where a producer is able to set market prices and maintain them above cost for a sustained period. It is based both on economic theory and on some evidence from experience which suggests that competition between many suppliers will drive price down towards cost while promoting innovation and quality (Geroski 1991). Where substitutes for particular goods or services exist, suppliers will find it difficult to market products which are inferior to or more costly than the substitute. Where there are no substitutes for the good, the producer is in a market by itself – a monopoly – and is potentially able to earn supernormal profits. To prevent this, competition policy seeks to encourage the provision of substitutes by removing barriers to market entry.

The main insight of recent literature on competition policy is to question the previously dominant assumption that structure determines behaviour. Geroski (1991) shows that, statistically, variations in industry structure account for only a minority of inter-industry variations in outcome. Industry structure makes certain types of anti-competitive behaviour possible, but whether and to what extent these materialize depends on strategic behaviour. In sum, industry structure shapes the options open to firms, but firms' decisions can also shape industry structure. A concentrated industry structure is only *prima facie* evidence of anti-competitive behaviour. Hence, as London Economics (1994a) and NERA (1992) suggest in papers for the Office of Fair Trading (OFT), public policy should shift to identifying and preventing the behaviour itself, where possible, or the potential for such behaviour where not (e.g. in a take-over). The OFT papers provide two relevant insights into how to diagnose and remedy such undesirable behaviour.

First, they show market structure is only a temporary guide to whether anti-competitive behaviour will be possible. In the medium term, structure reduces to whether new competitors can enter the market. Barriers to entry and strategic behaviour shape industry structure in the long run. Affecting those factors is the real remedy to concentration, and (therefore) to anti-competitive behaviour.

The second insight is the importance of market definition. To identify barriers to entry to a market, we need to be clear about the boundaries of that market. That is, we need to see if there are substitutes to an offending firm's products. This is usually done by measuring cross-price elasticities, that is showing how demand for good A varies with changes in the price of good B. We say that good B, say vodka, is a substitute for A, tequila, if sales of vodka increase when the price of tequila increases, all else being equal. By tracking movements in prices and revenues, evidence of sustained inelasticity can be found so as to identify non-competitive markets.

Applied to the media, this approach would make possible a much clearer definition of markets, based on consumers rather than technology. For example, it would point up the existence of separate markets for media goods (sold to consumers) and for media airtime and advertising space (sold to other producers). Within consumer markets, it would base its definitions on consumer use rather than technological difference: for example, to the extent to which satellite movie channels are a substitute for video rental, Blockbuster may be thought to be in the same market as Sky Movies.

Competition Policy in the UK

The UK is the only EU country, apart from The Netherlands, not to have incorporated European anti-competitive provisions into law (*Financial*

Times 5 May 1995 p. 10). Articles 85–90 of the Treaty of Rome establish the competition framework of the European Union, but they are concerned with instances of anti-competitive behaviour which affect cross-border trade. UK competition policy lacks comprehensiveness and does not impose penalties for anti-competitive behaviour (see Sir Brian Carsberg's arguments in *Financial Times* 24 February 1995 p. 15). European anti-trust laws impose penalties on firms that fail to supply information, when a mandatory request is issued. From the EU perspective, firms have a duty to co-operate with the investigating authorities. UK arrangements fare very poorly in comparison.

Effective, national competition regulation should be the primary plank of media policy. European competition legislation should be incorporated into British law, with thresholds that would allow investigations into dominance of national markets. A prohibitory regime, with real investigative powers, would serve its objectives more effectively than the current administrative-type system. Such a regime, enforced by fines (the European Union is empowered to impose fines of up to 10 per cent of the world-wide turnover of offending firms) and the power to reverse offensive actions, would improve the UK's competition laws and serve the interests of the public and of consumers. But strengthening UK competition law will not be sufficient to deal with legitimate public concerns about the structure and performance of the UK media. That is why ceilings on ownership concentration are needed.

Competition and pluralism: two birds with one stone?

Useful and important though competition policy is, it is dangerous to assume that it is an 'open sesame'. As David Aitman says: 'The goals of the maintenance of pluralism and the maintenance of competition are, at times, convergent; however, this is not always so and it is in these cases that different approaches are necessary' (1994 p. 16).

In order to see where competition policy cannot hit both the competition and the pluralism birds with one stone, it is worth sketching the main differences between achieving competition and pluralism.

Well grounded in theory and precedent, competition policy promises to promote pluralism (many providers) and diversity and quality (differentiation between products and customer oriented services). The theoretical potential of competitive markets to respond to demand signals from consumers, and to engender pluralism, diversity and quality without political intervention, explains why the application of competition policy principles to the marketplace for ideas has seemed so attractive. We can anticipate some disadvantages to regulating media solely through the application of competition policy.

First, there may be a discontinuity between competition policy's emphasis on homogeneity and the democratic requirement of diversity. Media products are inherently difficult to substitute for each other. How far are the *Sun* and the *Financial Times* mutually substitutable? The essence of media markets is the attempt to develop unique products for which there can be no satisfactory substitute. True, there is some degree of substitutability in media markets. Channel 4's success in broadcasting 'Football Italia' as the English Premier League moved to pay TV testifies to this. But the lesson to be drawn from the most recent experiment[1] in testing the limits of UK national newspaper markets and the degree to which one newspaper is a substitute for another is that the national daily market is better considered as made up of distinct markets rather than being a single market. Media markets are heterogeneous, whereas competition policy is more conveniently applicable to homogeneous markets.

Second, it is possible to foresee cases where competition policy might tolerate abuses of pluralism. Competition law will judge whether a company is dominant in any one economic market. It would not consider whether a portfolio that is not economically dominant in any one media market could add up to an abuse of pluralism across all media markets. Conversely, competition policy could tolerate a huge, efficient provider in a contestable market with no barriers to entry (Baumol and Sidak 1994). The correct definition of this relevant market would include companies not currently making the product or offering the service in question but which could switch to doing so (London Economics 1994a; Dertouzos and Trautman 1990), and would result in such a market being competitive but not characterized by diversity or pluralism. And, even though the Monopolies and Mergers Commission has specific responsibilities in respect of the media (as specified in the Broadcasting Act 1990, Part X, Section 192, referring to the Fair Trading Act 1973), operation of these provisions has been described as 'ineffectual' (Curran and Seaton 1991 p. 286).

Third, many doubt that competitive media markets, even those with more than one firm supplying relevant products and services, do deliver diverse, quality media products. They characteristically point to broadcast television in the USA where there is significant convergence of programming content. Blumler found that the US terrestrial television regime was 'inimical to broadcasting range – of programme form, of sorts of quality favoured, of viewer effect and experience stimulated' (1986 p. 141). The simultaneous screening of the O. J. Simpson trial on seven separate channels in the United States has done little to challenge Blumler's judgement. Economic theory names this phenomenon 'Hotelling's effect'.

Hotelling's effect applies under special circumstances, namely where there is non-price competition (e.g. between advertising funded television channels). Economically rational competitors will then crowd in the middle of the spectrum of consumer tastes rather than provide a diverse range of products. Assuming consumer tastes can be arranged in a continuum, it can be argued that two firms that did not compete on price would do best by positioning their products 'where the demand is', in the middle of the continuum.[2] New entrants would face the same incentives, re-sulting in an undue tendency for competitors to imitate each other. Advertising funded, profit maximizing television is a case in point. There is no price nexus between final consumers and suppliers and it is widely acknowledged that programming on advertising financed US television is driven by the needs and interests of advertisers rather than by final consumers.

Hotelling's effect may hold where there are a limited number of competitors but is less convincing where there are many competitors. There has to be a point after which it is more rational to pitch products and services to a market niche than to intensify already hot competition in the middle of the road. This analysis helps explain the parallel schedules and programmes that BBC1 and ITV have evolved, but will be less useful in a world of tens, if not hundreds, of media channels, competing on price as much as on non-price criteria. Applying Hotelling's effect to new media suggests that the middle ground of each niche market may be crowded, but not that an explosion of bandwidth will result in hundreds of identical general entertainment channels. However, Hotelling's effect does reinforce the recognition that competition will not necessarily guarantee diversity in the range and character of products available to consumers.

There is a further difference between media and other markets. Whereas in markets for other products it may be perfectly acceptable for one supplier to go out of business and consumers be served by others offering close substitutes for the products no longer available, cessation of publication of a newspaper title may deprive the readers of that title of a product which they value highly and for which they regard substitutes as unacceptable. Moreover, the loss of a voice articulating a particular point of view may be seen as impoverishing society as a whole (see, *inter alia*, Curran and Seaton 1991 pp. 107–8).

In sum, competition policy is necessary but not sufficient to deliver a democratic marketplace for ideas and to provide a comprehensive basis for the regulation of media markets. It is an important tool because:

- The media is a significant economic sector, and therefore we need competition policy to keep prices and costs down and quality up. Implementing such a policy may involve specific difficulties, but not ones that would be insurmountable to a media-specific regulator.

- Competition policy may not be sufficient to regulate the marketplace for ideas, but it helps. Open entry and competition between rival sources will go some way to meeting our democratic aims: ensuring that consumers can choose and test ideas from and against a wide range of sources.

However competition policy lacks sensitivity to problems posed by media enterprises which compete in several markets but are dominant in none. Competition cannot therefore solve all the problems in UK media markets. And it should be supplemented by regulation to promote pluralism, diversity, impartiality and accuracy.

Plurality, diversity, impartiality, accuracy

Little conceptual clarification has been attempted in respect of plurality, diversity, impartiality and accuracy – the building blocks of a democratic media. Whereas competition policy can identify and measure behaviour that is anti-competitive, democratic policy has no comparable conceptual framework. The link between pluralism (structure) and accuracy, diversity, impartiality (behaviour) is not straightforward. For example some have reasoned from structure to behaviour to argue that the BBC's 19.7 per cent share of the UK media market (Sanchez-Tabernero 1993) constitutes a flagrant offence to pluralism. Judged on one calculation of 'national share of voice' the BBC has a share of voice 'nearly twice the weight of its nearest rival' (*Financial Times* 21 March 1995 p. 11). Yet few have argued that the BBC's behaviour is offensive to the principles of accuracy, diversity and impartiality.

The share of voice index created by the British Media Industry Group (1995) purports to measure the impact of different media on consumers by aggregating the share of consumption, and thus putatively of influence, of different UK media enterprises. It aggregates regional and national newspaper circulation, radio listening and television viewing. It weights radio differently, notably by discounting its share, and notionally therefore its impact, by 50 per cent. The index attributes a 19.7 per cent share of voice to the BBC and a share of 10.6 per cent to News International. All other UK media enterprises are stated to enjoy less than a 10 per cent share of voice (p. 11).

Whether or not the concept of share of voice is meaningful, the index is useful in focusing attention on the unexamined assumption that structure is a satisfactory proxy for behaviour. The assumption is pervasive and possibly well founded, but unproved. No research has demonstrated continuities in content between media in common ownership and discontinuities between them and other media in other ownership. It may be legitimate, and politically possible, to legislate to avoid possible anticipated future harm. It is less obviously legitimate, and may be politically impossible, to act upon lawful actions by firms on the basis of anticipation of a possible future harm.

The case of the BBC makes clear these difficulties. If a share of voice methodology[3] were to be adopted as a basis for regulation of cross-ownership or concentration of ownership then, *prima facie*, the BBC is first in the firing line. However, no harm arising from the BBC's putative dominant position in share of voice and share of reception has been demonstrated. Indeed many would argue (and we would agree) that the BBC's behaviour makes its share tolerable. Like all organizations, the BBC's achievement of its goals, notably those of accuracy, impartiality and diversity of content, is imperfect. But, its performance is sufficiently good for its notionally dominant share of voice and audience to be accepted. This is not to say that the editorial independence of BBC journalists should not be strengthened and the BBC's accountability to its users should not be improved (see chapter 7) but only that the BBC's performance has been sufficient to justify tolerance of its share of voice and audience. Moreover, there are occasions when pursuit of the goal of diverse ownership is likely to reduce pluralism of outlets – for example, if newspapers were to close because their acquisition was blocked to prevent concentration of ownership. This is a particularly difficult issue. If a bid from a dominant firm for a failing title or broadcaster were to be disallowed on the grounds of increasing dominance, the failure of the title would still leave us with fewer sources and a greater concentration of ownership. Liberty acknowledges that News International's acquisitions of *Today* and *The Times* fit into this category (1994 p. 51). Indeed the closure of News International's *Today* title in 1995 was presented as a response to the prospect of more stringent regulation of media ownership. The relationship between pluralism of ownership, pluralism of outlets and diversity and reliability of opinion or content is not straightforward.

If (as we suggest arguing from the case of the BBC) some types of behaviour make dominant shares of media markets tolerable and links between ownership, behaviour and content are opaque, achievement of public policy goals of diversity, accuracy and impartiality in media content is best pursued through a combination of means. Public funding for media, guarantees of editorial independence and content rules should be given higher prominence in an overall portfolio of measures, including limits on concentration of ownership, to better ensure diversity and reliability of information.

But that is not to say we should tolerate abuses of plurality, just because we are taking other measures to foster diversity and reliability of information. Abuses of pluralism directly threaten democracy: we would not tolerate our media industries being dominated by a single company, however diverse its products. Even if *Pravda* and *Izvestia* had been sources of diverse, reliable news, pluralism in the Soviet Union would have been limited and democracy hindered.

An effective competition policy would remedy or prevent some cases of concentration, but ownership limits are needed as the braces to guard against those cases that slip through the belt of competition. These pluralism braces should be narrowly targeted at industry structure. They should define the marketplace for ideas. They should respond to the question: what is the minimum number of media companies we would tolerate? They should then fix a ceiling that reflects that agreed minimum. Those ownership limits would trump competition policy, would be technology neutral and sensitive to share.

Cross-ownership

Safeguarding pluralism is the reason for ownership limits. How does the current policy framework measure up? It has two distinct characteristics:

- cross-ownership limits, or limits to ownership across different media markets
- limits to concentration within any one media market.

Cross-ownership is generally assumed to be contrary to the public interest. In 1962, the Pilkington Committee stated that where the ownership of radio, television and/or newspapers is combined in a single firm, democracy is threatened by 'such an excessive concentration of power to influence and persuade public opinion [which might be used for] a one-sided presentation of affairs of public concern' (p. 182). The Committee's concern resulted in the regulator of commercial broadcasting, then the Independent Broadcasting Authority (IBA), prohibiting newspaper proprietors from holding a share in an ITV programme contractor where such a holding would be contrary to the public interest. In practice, the IBA exercised discretion in implementing this vague, but clearly cautious, approach to the regulation of cross-ownership.

Gibbons (1991) reports two opposing criticisms of the way the IBA interpreted this 'fuzzy' brief. The IBA was criticized for its vulnerability to commercial pressure in respect of Rupert Murdoch's shareholding in LWT during the late 1960s and early 1970s. Murdoch acquired his stake in LWT despite holding shares in a London commercial radio station and in newspapers circulating in London. Following criticism, the IBA inter-vened and effectively froze the ownership of ITV licences: ownership could only change during franchise rounds.[4] Its stance exposed it to criti-cism from an opposing viewpoint: companies holding ITV franchises were now too insulated from commercial pressure.

The Broadcasting Act 1990 attempted to deal with both criticisms. Schedule 2 of the Act details the limits to be placed on cross-ownership, removing the regulator's discretion and therefore its vulnerability to commercial pressure. However, in attempting to predict and prevent all loopholes in advance, Schedule 2 became overly complicated and detailed, and the Act as a whole is one of the longest in the statute book. After four years of accelerating change in the media, the provisions of the Broadcasting Act 1990 have come to look increasingly inflexible and anachronistic.

By May 1995, limits based on the number and type of channels or newspapers were creaking. Whatever their combined share, it was pos-sible to lawfully control a national newspaper and an unlimited number of satellite channels, but not a national newspaper and an ITV licence: the *Guardian* could not have acquired Border Television, even though the resulting share of the newspaper and television markets would be dwarfed by that of News International's portfolio of BSkyB and five national newspapers. Moreover, there seemed to be no obvious equiva-lence between the levels embodied in the Broadcasting Act 1990 and those in the Fair Trading Act 1973 which provides that any transfer of newspaper ownership of titles whose circulations exceed 500,000 copies requires the consent of the Secretary of State.

These contradictions suggested that the government's consideration of cross-ownership (as distinct from concentration of ownership) rules was timely. After months of contradictory briefings, the government pub-lished its proposals in May 1995 (Department of National Heritage 1995). Its short term proposals will ease cross-ownership restrictions while attempting to maintain ceilings on concentration of ownership. The long term proposals for a share of voice system would abolish cross-ownership rules altogether. Broadly, the government's proposals are sens-ible and, given the favourable reception they received, we must evaluate whether the cross-ownership rules which remain are needed in the long term. To do so, it is crucial to understand that *cross-ownership* rules were introduced to inhibit *concentration* of ownership. They were established because of concern over concentration of ownership rather than because cross-ownership was a cause for concern in itself. They were the means taken to achieve the end of limiting concentration. We must be careful not to confuse means and ends: concentration of ownership matters; cross-ownership, in itself, does not.

In the old media world, ownership of both television and newspapers was, almost of necessity, a threat to pluralism because there were few products in each market. In the new media world more products and services are becoming available and the audience for individual titles and channels is fragmenting. True, this tendency has not resulted in a plethora of uniformly sized media companies. True, therefore, that cross-ownership rules continue to prevent some groupings which threaten pluralism. The point is that they also prevent some groupings that do not threaten pluralism, while allowing others that might. Less and less will cross-ownership be a convenient proxy for concentration, so we must find a system that directly measures and prevents threats to pluralism. We require a new means to achieve the same end – reduction of the potential for harm to the political process presented by concentration of media power.

Concentration of ownership

This section evaluates current UK policy on media concentration in the light of the policy considerations discussed above. The ideal pluralism policy would:

- define the marketplace for ideas
- decide the minimum tolerable number of media owners
- fix an ownership ceiling accordingly.

The ideal pluralism policy would therefore be technology neutral and sensitive to share. The current UK regime has the following characteristics.

First, for newspapers, the Fair Trading Act 1973 is the principal basis for action. Section 58 requires the consent of the Secretary of State for Trade and Industry for any acquisition which will result in ownership of newspapers with a cumulative circulation of over 500,000. It is a criminal offence to proceed with a newspaper merger without this consent. The Secretary of State's consent to such an acquisition must normally follow referral of the merger proposal to the Newspaper Panel of the Monopolies and Mergers Commission. Exceptionally the Secretary of State is empowered not to refer to the Monopolies and Mergers Commission if the newspaper(s) involved in the transfer is (are) not viable as a going concern(s). In such a case consent to the acquisition must be given unconditionally, without requiring a report. Where the Monopolies and Mergers Commission is required to report the Commission must determine whether the acquisition would be in the public interest, taking into account 'in particular, the need for accurate presentation of news and free expression of opinion'.[5]

Second, for television, the rules governing concentration of ownership before the government's recent announcement are to be found in Section 5 and Schedule 2 of the Broadcasting Act 1990. In respect of ITV, the intention of the 1990 Act was to protect ITV's federal, regional ownership structure, which had been created by the ITA after the Television Act 'to prohibit the creation of a centralised monopoly based on London' (Curran and Seaton 1991 p. 215). The 1990 Act and Schedule 2 laid down that a maximum of two licences could be held by any one corporate body. In addition, it divided the licence regions into large and small, and prohibited common ownership of two large licences. With Peter Brooke's (then Secretary of State for National Heritage) revisions to Schedule 2 in 1993, the government effectively abandoned its attempt to underpin, with regional ownership, the regional structure of ITV. Two large licences could be owned simultaneously and three major mergers followed in short order: between the two largest licensees, the third and fourth largest, and the fifth and sixth largest (respectively, Carlton-Central, Granada-LWT, MAI-Anglia). These changes were made by statutory instrument: the greater flexibility this allows appears to have been bought at the price of greater government involvement.

How do these arrangements measure up to our criteria? Not well. There is no definition of the marketplace for ideas. There is no justification given for the ownership limits chosen. Both systems are technology-specific, and the Broadcasting Act 1990 is not sensitive to share.

There are clear anomalies: Granada is able to own LWT, but LWT would not have been able to buy Yorkshire-Tyne Tees, because it would then have controlled three licences. Yet, Granada-LWT has a larger share of the ITV advertising market than LWT-Yorkshire-Tyne Tees would have had. Had LWT been able to bid for Yorkshire-Tyne Tees, the outcome of the LWT take-over battle might have been different. The ITV structure was reshaped not by commercial or pluralist logic, i.e. not by the public interest, but by the accident of phrasing of a statutory instrument.

These rules were not sensitive to market share. They were not technology neutral. They were overly complicated. They distorted competition but did not prevent abuses of pluralism. To avoid the anomalies they created, we need a system that has a clear rationale, implemented flexibly, bound neither to technological distinctions nor to numbers or types of media outlet.

The government's short term proposals on media ownership (Department of National Heritage 1995) take a step back in terms of complexity, but two steps forward in terms of share sensitivity and technology neutrality. They allow newspapers groups with less than 20 per cent of their market to own no more than 15 per cent of each of the television and radio markets. Previously, newspaper groups had been banned from controlling terrestrial television licences, and vice versa. This clearly increases sensitivity to share and technology neutrality. But by overlaying a system based on numbers and types of licences with one based on share of market, the proposals create a more complicated hybrid. While welcome, these short term improvements do not put off the need for a comprehensive solution to the problem.

What would such a comprehensive solution look like? Competition policy will do much of the work: we need rules on concentration of ownership to do what competition policy cannot. Remember that these rules are not aimed at diversity: they are narrowly targeted at pluralism, at preventing anyone controlling too great a share of the media. Diversity is a matter for competition policy and content regulation.

It is impossible to define objectively and precisely what 'too great a share' is and therefore where ownership limits should bite. The British Media Industry Group (1995) and Andersen Consulting (Shew 1994) have each suggested a means of calculating shares. But neither says what too great a share would be: creating the ruler does not tell us what we are measuring. Should it be share of revenue, share of consumption or what? Should consumption of one medium be weighted as more important than share of another? We need to say what too great a share is. Once we admit there is no objective answer, this becomes easier to do. Better to ask the question: what is the lowest number of media owners tolerable in the UK? We believe that a market of ten media owners does not threaten pluralism, but that one of five would.

If ten owners, each with no more than 10 per cent of the market, are sufficient to ensure a satisfactory level of pluralism, and five are too few, then no less than seven owners, each with no more than 15 per cent of the total media market, seems a reasonable definition of a floor for ownership regulation. But 15 per cent of what?

What and how we measure is crucial: it can mean the difference between Carlton having a 3 per cent or 12 per cent share. It can make News International have 3 per cent or 22 per cent. Most media players are proposing share of voice models. The weights assigned to different media and to public broadcasting determine the outcome. The greater the size of the market, and the smaller the weight assigned to one's own company, the more permissive the result. This extreme sensitivity to subjective assumptions raises doubts about the objectivity and acceptability of any share of voice system. Even if those doubts were overcome, it would be difficult to implement: lots of information would have to be gathered, there might be perverse incentives and no single way of measuring market share would be universally applicable.[6] In short, the system would be open to political and commercial pressure: it would be neither fair nor seen to be fair.

A more robust solution exists. Allow companies a 40 per cent ceiling in not more than one of several distinct markets. But if they have a share of more than 5 per cent in an additional sector, the amount they can own in each sector falls, as indicated in table 4.

Which four sectors? National newspapers, regional newspapers, television and radio, because they are the main sources of news and opinion. Any firm controlling more than 15 per cent of the aggregate would have too much influence. Magazines are omitted because they have low barriers to entry and are less important for news and opinion, and the magazine market demonstrates ample pluralism.

What is the basis for the numbers? The rationale is the minimum tolerable number of owners, mentioned above. This gives us a 15 per cent market share limit in each sector for anyone present in all four sectors. That limit is then relaxed progressively if a proprietor is not present in all four sectors. The 40 per cent criterion is taken from the European Commission (Aitman 1994).

How are the shares measured? National and regional newspaper shares are measured by circulation figures. Free-to-air television and radio shares are measured by the BARB/RAJAR statistics and, in default of these, by a company's declared advertising audience. Subscription television shares are measured by number of subscribers. Each sector can be measured in the most appropriate way, precluding the need for an elusive cross-media yardstick.

Why include regional papers in the national figures? Because regional papers are a part of the country's climate of opinion.

This system would be measured nationally and would protect national pluralism. But it could not so easily be applied to regional media, because there is often only one regional paper or television station in any one regional market. A system that is sensitive to share is not yet appropriate to regional markets. Instead, we propose to preserve the principle embodied in the Broadcasting Act 1990: that there should be no double monopoly in any region.

These rules are based on judgements. But so are and will be all other systems. At least these rules would be clear and flexible without being destabilizing, insensitive to share or overly sensitive to technology.

Table 4

Maximum share in *one* of the following sectors:	Maximum share in *two* of the following sectors:	Maximum share in *three* of the following sectors:	Maximum share in *each* of the following sectors:
• national newspapers • regional newspapers • national television • national radio	• national newspapers • regional newspapers • national television • national radio	• national newspapers • regional newspapers • national television • national radio	• national newspapers • regional newspapers • national television • national radio
40%	30%	20%	15%

Behaviour, not structure

Given our conclusion that limiting ownership is but one of the necessary policy tools, surprisingly little attention has been given to policies that further diversity and reliability of information. Little has been written about journalistic and editorial independence since the Royal Commission on the Press's chapter (interestingly the shortest of the Commission's report) on editorial contracts (McGregor 1977 pp. 154–6). Given the Commission's rhetorical emphasis on the relationship between proprietors and editors, subsequent neglect of the matter is particularly surprising. The Royal Commission found: 'the importance of publicly and explicitly guaranteeing editorial independence, in the context of concentration, chain ownership and monopoly ... is so great that publishers and editors should urgently consider ways of setting out and guaranteeing formally editorial independence from employers' (p. 155). We agree.

The complexities of establishing a robust and equitable basis for regulation of concentration of ownership of the media in the UK are apparent from our discussion above. Given these difficulties (and the problematic relationship between structure of firms and markets and the behaviour of owners identified by Geroski 1991) it is worth asking whether the goals not sought through ownership regulation (freedom of expression, diversity of opinion and accurate reporting of events) can be achieved, in whole or part, by other means. It is clear that none of these goals can be achieved through regulation of the ownership of the media alone. For they concern the behaviour of media workers, on whom the quality and character of media content depend, and the behaviour of journalists and editors cannot be 'read off' from the structure and ownership of firms. Professional values enshrined in jealously guarded traditions of investigative reporting, iconoclastic judgement and editorial independence are necessary, if not sufficient, conditions of healthy media and a healthy democratic polity. How can they be strengthened?

The Royal Commission defined the basic rights of editorial and journalistic independence as:

- the right to reject material provided by the central management or editorial services
- the right to determine the contents of the paper (within the bounds of reasonable economic consideration and the established policy of publication)
- the right to allocate expenditure within a budget
- the right to carry out investigative journalism
- the right to reject advice on editorial policy

- the right to criticize the paper's own group or other parts of the same corporate organization
- the right to change the alignment or views of the paper on specific issues within its agreed editorial policy
- the right to appoint or dismiss journalists and to decide the terms of their contracts of employment within the established policy of the organization, and the right to assign journalists to stories (McGregor 1977 p. 155).

It acknowledged the difficulty of defining the limits of the proper responsibility of editor and proprietor, and noted that 'editorial and managerial decisions were inseparable' (p. 155) and that, therefore, the editorial policy could not and cannot be separated from the financial success of a media enterprise. None the less, we believe that an explicit codification of editorial and journalistic independence along the lines of the catalogue of 'rights' defined by the Royal Commission is necessary if the goal of pluralistic, independent and investigative media is to be achieved. In the first instance we believe that consent for further mergers and concentration of ownership of the media in the UK should be conditional on a strengthening and extension of editorial and journalistic independence. We make this recommendation because we recognize that regulators may, again, be faced with the invidious choice between allowing a media outlet to die, resulting in increased concentration of ownership, or be taken over by an established dominant player, also resulting in an increased concentration of ownership. The harm might be minimized by permitting a merger on condition that editorial and journalistic independence were strengthened.

The Royal Commission concerned itself with editorial and not journalistic independence. Indeed, it saw editorial independence threatened as much by journalists as by proprietors. Accordingly it was unsympathetic to proposals for control by workers and journalists. Ascherson, in a response to the Royal Commission, argued for what he euphemistically dubbed 'internal democracy' (1978 p. 124) in the press, albeit he acknowledged that the cause of 'internal democracy' had been weakened by the fate of the *Scottish Daily News*. His commentary preceded the mayfly life and death of the *News on Sunday* but the fate of that newspaper can only have further weakened Ascherson's cause. None the less, it seems wrong to follow the Royal Commission and limit 'rights' of independence from proprietors to editors, and accordingly we propose that such rights be enjoyed by journalists as well as editors and be embodied in any agreement of new rules on concentration of ownership of the UK media, which must be conditional on an entrenched independence for the editors and journalists concerned.

Summary

1 Large, concentrated media organizations are not intrinsically undesirable. Large size tends to bring the resources required for comprehensive high quality reporting and the case of the BBC suggests that large organizations with a high share of media markets can serve the public interest.

2 The link between structure and behaviour is not straightforward. Policies promoting structural change, promoting pluralism in ownership, may be inimical to diversity of products and services.

3 Different policies may be required to foster pluralism and diversity.

4 Significant harms may arise from the dominance of a single organization in a distinct media market or across several different markets. Accordingly regulatory measures are necessary to prevent firms abusing a dominant position.

5 Assessment of the extent of firms' dominance in particular markets, and across several markets, is highly sensitive to how relevant markets are defined.

6 Cross-ownership is in itself not significant; what counts is concentration of ownership.

7 Three remedies should be adopted:

  (a) Stronger and more comprehensive competition laws, based on European Union precedents and on Articles 85–90 of the Treaty of Rome and the Council Regulation on the control of concentrations between undertakings (Council Regulation 4064/89 of 21 December 1989).

  (b) Measures to strengthen journalistic and editorial independence and thereby weaken any link between ownership and control of media.

  (c) Sector-specific limits on concentration of media ownership in four sectors: television, national newspapers, regional newspapers and radio. Media ownership should be limited to a single firm having no more than a 40 per cent share of any single media market, no more than 30 per cent of each of two media markets, no more than 20 per cent of any three media markets and no more than 15 per cent of each of four media markets.

4 Universal Service Obligation in Broadcasting and Telecommunications

Universal service obligations (USOs) are rules imposed on the suppliers of services identified as part of people's basic entitlements to media and communications to ensure that they are available to all at affordable cost. In telecommunications the universal service obligation has customarily been defined as universal access to voice telephony at affordable cost. In broadcasting the universal service obligation has seldom been defined as such, but a broad consensus exists that freedom of access to the information necessary to full participation in economic, political and social life is a central element of citizens' entitlements in modern societies.

Most developed countries have extensive infrastructures and near-universal penetration of key services so the issues of USO may appear irrelevant. Not so. USO issues are in the forefront of public policy because technological and organizational changes to broadcasting and telecommunications in recent years have meant that the established means of achieving this end have been put in question. First, both telecommunications and broadcasting have evolved from a state owned monopolistic structure to a more complex mix of public and private operators. The general provision to supply 'anyone who requests service' imposed on the incumbent player in the market, with no attention to the costs of doing so or the efficiency with which this task is discharged, is no longer adequate in a competitive market. The publicly owned monopoly which formerly undertook delivery of USOs is a monopoly no longer (and in telecoms is no longer publicly owned). Not only must policy makers decide how USOs should be defined and which firms should shoulder the responsibility of discharging them but they must also determine how USOs should be paid for. Second, the social and technical characteristics of both telecommunications and broadcasting markets have changed. Reappraisal of USOs and the means by which they are to be achieved is urgent. We examine universal service obligations in broadcasting and telecommunications in turn.

USOs in broadcasting

Universal service at affordable cost is one, albeit the most solidly enshrined, of the user entitlements in media and communications. Although this formulation has customarily been adopted for telecommunications it also characterizes important aspects of public service broadcasting. But it does not provide a comprehensive definition of citizens' entitlements in broadcasting. The right to freedom of information and expression, which is defined in a variety of international agreements including the European Convention on Human Rights, must also form part of the USO. Article 10 states: 'Everyone has the right to freedom of expression. This right shall include freedom to hold opinions and to receive and impart information and ideas without interference by public authority and regardless of frontiers'.

Yet access to services and freedom of information and expression do not capture all broadcasting entitlements. Scannell offers a more inclusive definition:

> Communicative entitlements presuppose communicative rights. Communicative rights (the right to speak freely, for instance) are enshrined in the written constitutions of some countries but not in Britain. A minimal notion of guaranteed communicative rights is a precondition of forms of democratic life in public and private . . . Communicative entitlements can be claimed and asserted, within a presupposed framework of communicative rights. Rights of free assembly, to speak freely and (more often overlooked) to listen, contribute to creating formal, minimal guarantees for certain forms of public political and religious life. They seed the possible growth of wider and more pervasive claims from those denied a hearing in manifold public, and private contexts, that they should be listened to: i.e. that they should be treated seriously. As equals. (1989 p. 160)

This and similar definitions of a basic communicative entitlement seem unexceptionable: Scannell points towards a duty of ensuring the expression of a plurality of voices and promotion of opportunities for them to be heard. But viewer and listener rights and entitlements are often defined even more comprehensively. The collective cultural rights of viewers and listeners too are part of citizens' entitlements and have

formed a basis for regulation of the proportion of foreign content transmitted in any particular medium and for initiatives such as the Welsh
Fourth Channel, S4C, and the Gaelic Programme Fund.

Cultural rights

Jacques Delors, when President of the Commission of the European
Communities, stated that:

> Culture is not a piece of merchandise like any other and must not be
> treated as such ... culture cannot flower today unless control of the relev-
> ant technologies is assured. On the first point ... we cannot treat culture
> the way we treat refrigerators or even cars. *Laissez-faire*, leaving market
> forces to operate freely, is not enough. I would like to ask just one ques-
> tion of our American friends . . . do we have the right to exist? (Assises
> européennes de l'audiovisuel 1989 pp. 47–8)[1]

Here, Delors implicitly constitutes collective cultural identity as a *right*.
It can go without saying how important his views are – both because of
the powerful position he long filled with distinction and because of the
representative character of his arguments.

How are we to interpret a right to cultural identity? According to
Taylor (1993) the modern conception of rights revolves around the core
idea of respect of human integrity. In turn, respect for our integrity presupposes that the conditions for our identity are respected. Hence the
right to collective identity is the entitlement to:

> certain values, certain allegiances, a certain community perhaps outside of
> which I could not function as a fully human subject ... what is peculiar to
> a human subject is the ability to ask and answer questions about what
> really matters, what is of the highest value, what is truly significant ... the
> conception of identity is the view that outside the horizon provided by
> some master value or some allegiance or some community membership, I
> would be crucially crippled, would be unable to ask and answer these
> questions effectively, and would thus be unable to function as a full human
> subject. (1993 p. 45)

The UK has attempted to secure expression of collective identity and a
range of voices by creating institutions dedicated to the achievement of
this goal, the public service broadcasters, notably the BBC and Channel
4; by requiring commercial broadcasters to serve these goals as a condition of licence; and by subsidizing production in minority languages, in
Welsh via S4C and in Gaelic via the Gaelic Programme Fund. But the
sustainability of these initiatives is likely to decline. Trading off of

programme quality and character against the right to use the radio frequency spectrum and enjoy an advertising monopoly is a bargain governments are no longer able to enforce. National broadcasting regulation is no longer able to guarantee exclusive rights to the sale of television advertising in particular markets. And radio frequencies suitable for broadcasting have become less scarce – a process which is likely to continue. Accordingly, other methods of securing the entitlements to which Scannell refers must be found.

Few governments will raise taxes to sustain increasingly expensive public broadcasting services. Some countries allow the public broadcasters to complement their revenues with advertising. Others have sought new ways to realize citizens' communication entitlements and have supplemented public service broadcasting services with programme production funds for which rival firms bid competitively. New Zealand is the most striking example of this initiative and Canada's Telefilm production fund is another case in point.

In the UK, the 'community service obligation' in broadcasting – a term borrowed from Australian regulation[2] – has taken two forms: first, *universal accessibility* of services which are free at the point of use; and second, ensuring that programmes satisfy what the ITC has called *positive programme requirements*.[3] Of course, programmes must also satisfy 'negative' requirements: they must satisfy taste and decency requirements, advertisements must conform to the ITC's advertising code, and so on.

The need for an obligation to provide universal reach arises from the fact that some viewers and listeners could never be economically served, despite the fact that advertising supported television gains from maximizing reach. Those living in small and remote communities are disproportionately expensive to reach and low income households are unattractive to advertisers. The striking success of subscription television has demonstrated that the ability to *exclude* non-paying consumers and focus on specific programme demand (for sports or films) yields high returns. Moreover, the costs of extending a service to remote communities are set to grow: digital terrestrial distribution is more expensive than analogue terrestrial transmission. Accordingly, without regulatory intervention the broadcasting market is likely to provide *fewer* radio and television programmes and services free at the point of use. Regulatory intervention is also likely to be required to ensure that the range and character of services will enable viewers and listeners to participate fully in social and political life.

The community service obligation of broadcasting can thus be defined as viewers' and listeners' entitlement to access to a range of information, entertainment and educational programmes at affordable cost.[4]

The continued existence of strong services, universally available and free at the point of use, will provide both a floor of quality below which competing commercial services will fall only at their peril and an incentive to innovation in new services. 'Me too' television and radio are unlikely to succeed if services like the current portfolio of terrestrial services remain available. We therefore believe that no fewer than five terrestrial television channels (BBC1, BBC2, ITV, Channel 4/S4C, Channel 5), and publicly funded national and local radio services, should continue to be available to all UK viewers and listeners free at the point of use. The continued imposition of positive requirements on the five 'public service' content providers is also a necessary part of the community service requirement. But should the community service obligation be confined to them? And how will the community service obligation be realized when government no longer disposes of monopoly power over transmission capacity and thus the ability to reach customers?

Why should advertising financed (and subscription) broadcasting adhere to positive and negative programme requirements when it can escape them through alternative means of reaching viewers and listeners?

In the mid-1990s we see only the beginnings of these problems. Terrestrial broadcasting is still attractive for advertising financed television and radio because no alternative distribution has comparable reach. Accordingly, the ITC is likely to be able to require delivery of positive and negative programme requirements at least until the end of the century and probably for some time beyond. But the old bargain will be less and less easy to strike. Once satellite penetration approaches terrestrial reach, the option of reaching UK customers from less regulated EU countries will become more and more attractive.

In principle the community service obligation obtains in all media. The public's interest in full participation in the political, economic, cultural and social life of our communities is not confined to cheap and easy access to broadcasting. In many respects this entitlement is well satisfied by commercial media in the existing UK media order. There are, for example, few grounds for public provision in the newspaper and periodical sector. What other country has five national newspapers owned by different proprietors and providing coverage of the quality of the *Financial Times, Guardian, Independent, Daily Telegraph* and *The Times*? Barriers to entry are relatively low, specialist publications abound, even some free sheets have innovated interestingly by serving geographical and interest communities. Radio too has low barriers to entry and opportunities for different interest groups to be served. The exacting challenges lie in television, on-line services and the new media. Here, the emergence of vertically integrated network and service providers, and proprietary control over gateways and bottlenecks in distribution and transmission systems and over key resources of content, are already causing concern. Liberalized media markets and pluralization of the media mean that established regulatory practice of tightly specifying and monitoring conditions of licence becomes difficult and expensive to implement. Internationalization of communications means that firms may evade onerous regulatory regimes by originating services from offshore locations.

We propose that realization of the community service obligation in broadcasting should be through:

- the continued existence of publicly funded services and notably the BBC
- must carry requirements for the services which conform to positive (and negative) programme requirements on alternative delivery systems
- the granting of priority access to networks of limited capacity to services which are free at the point of use.

Furthermore, we believe there may be an important role for a publicly financed programming fund at some future date. If in the future advertising financed terrestrial services decline in importance to the point that the benefits of access to terrestrial frequencies are outweighed by the costs of providing positive programming, then subsidies to commercial broadcasters or support for further public channels may be required. This could be achieved by reserving portions of auctionable spectrum to services which conform to positive (and negative) programme requirements, if the UK opts for a market in spectrum. There is ample justification today, however, for investing public money in on-line services to stimulate the market and to ensure that there is a strong presence of material which satisfies positive programming requirements in this domain.

USOs in telecommunications

Most people associate USOs in telecoms with a general obligation to provide a widespread national network coverage, and to ensure that citizens who live in rural areas are not disadvantaged and left at the margin of society. This concept is inherited from the past, when public provision of telecommunications was the norm in Europe. Public policy goals were part of the general remit of the operators, and they were seldom formally defined, leaving it to the individual administrations to interpret and implement specific policies. The operator had a commitment to pass all homes in the country rather than to connect any given percentage of households. Following liberalization of telecoms markets, and the transition from public to commercial provision, universal obligations had to be clearly stated and included in operators' licences. USO is today an umbrella term which includes a range of different policies responding to economic as well as social needs.

In UK telecommunications, access to the public network is sanctioned in the obligation 'to meet all reasonable demand for basic telephone service, including rural areas' (conditions 1 and 2 of BT's licence), which was eventually widened to cover implementation of a residential Light User Scheme (condition 24D). Other obligations include provision of special telephony services to the hearing impaired (conditions 31A, 32, 33), free emergency calls (condition 6), free directory services for the blind and disabled (condition 3) and provision of public phones (condition 11). This USO is currently imposed on BT for the whole of the United Kingdom except for the Hull area, covered by Kingston Communications.

Oftel defines USO as 'the requirement to provide consumers with direct access to a switched telephone network, and the ability to make and receive voice calls, at a reasonable price' (1994a p. 40). Today's policies are aimed at ensuring that neither income, nor disability, nor location prevents citizens from accessing the telephone network. Telephone penetration in the UK has reached 93 per cent, although this national figure hides substantial regional differences, as we will discuss later.

The definition of USOs is not simply an academic issue. The need for such an obligation stems from the recognition that some customers are uneconomic for a profit maximizing operator to serve. The definition of USO provides a line below which the operator would find it uneconomic to serve the customer. Several variables in the very definition of universal service will affect the point at which the line is drawn. These variables are as follows.

Definition of the services included in the obligation   Basic services are usually the object of USO. Until recently USO was interpreted as an obligation to make access to the telecommunications network available rather than actively providing telephony services to users. Today the emphasis is shifting towards actual provision of given services. These can be one way or two way voice telephony, data handling, broadband capacity etc. If USO is imposed for the simplest possible service, like receive-only voice telephony, relatively few users will be considered uneconomic to connect, once the revenues they generate (increased traffic in their direction) are taken into account. The more basic the service, the smaller the number of customers who would need a USO to be connected.

Efficiency of the providing operator   Efficient supply is a dynamic concept. Technologies change with time and each firm's production possibilities improve along a learning curve. Customers who were uneconomic to serve with a fixed link telephone, for example, can be profitably connected to a wireless network.

Enforcement mechanism   USO can be imposed on one or more operators. Because of the varying degree of efficiency which characterizes each PTO, the mechanism with which USO is imposed and the choice of operators charged with it may affect the costs of providing universal service. If the obligation is put out to tender to the lowest bidder, for example, USO costs are likely to be brought down by competitive bidding.

Level of network rollout   If the telecom network is nearly completed, the marginal costs of providing an extra line are minimal. Conversely, if network development is low, the same fixed costs are shared among a smaller number of users and additional connections might require a costly increase in overall network capacity.

Costing techniques   Identical services can appear more or less profitable depending on how their costs are measured and/or their revenues projected. What types of costs are taken into account? What time frame is assumed? The assumptions made will influence the definition and degree of onerousness of USO.

Pricing requirements   How is affordable (reasonable) price defined? The cost of USO will also change depending on whether different tariff structures are possible for different types of users and different locations.

Different policy dimensions of USO

USO is a response to needs of an economic and social nature. There are therefore a number of different dimensions to it, each a solution to a single policy concern. USO is used to maximize network extension, capture network externalities and fulfil social goals. Only by spelling out the different needs that USO is called to serve can we strike the appropriate balance between efficiency and equity in pricing and universal provision of telecom services.

The policy of ensuring extensive, or 'universal', geographical coverage is designed for the promotion of a ubiquitous rollout of the network, which would not occur at the same rate or to the same extent with a free market. It is a public policy requirement to have full development of the network and enable customers to be connected at their home. In a mature market, where network coverage has reached the majority of households, this requirement imposes a small burden on the industry. The bulk of investment has been sunk and additional connections add relatively little to the common costs. Conversely, in markets where network development is far from completion, this requirement can be very costly. Widespread access to basic voice telephony also satisfies the objective of security, in that it enables access to emergency services and to incoming calls, including the possibility of receiving warnings of anticipated potential harm. This policy aim should be monitored with given target penetration rates, which may be set on a regional as well as national basis, to ensure that regional differences are smoothed out.

The national penetration rate hides important regional differences. Figure 2 shows the penetration rate by region: the percentage of untelephoned households in the North is over twice as high as in the South East.

Offering residential services at geographically averaged prices responds to the policy aim of promoting fairness and reducing unjustified inequality (or in the language of the European Union, 'social exclusion') which arises from location and distance from the centre. This policy is increasingly called into question, as prices closely related to costs are the basis for effective competition. Service providers charged with geographical averaging are exposed to cherry-picking by competitors who are not so

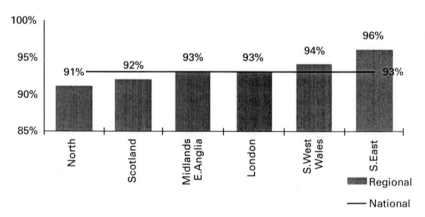

Figure 2   Telephone penetration levels by region
Source: Policy Studies Institute 1995

obligated. Furthermore, regional disparities in the prices of other utilities have shown that geographically specific pricing is feasible. Indeed the current system of cable franchises already results in tariffs that change from area to area. Abandonment of this policy would constitute a movement away from the goal of promoting social justice, and is likely to lead to the same (sparsely populated) areas being penalized, with all utilities relatively more expensive. We recommend that tariffs should stay uniform throughout each operator's licensed area, but not necessarily across different classes of customers, and that prices be brought in line with costs by more flexible and creative tariffing.

Installing and maintaining pay-phones in most public locations offer an important, although arguably inferior, substitute for universal residential provision. A portion of the total traffic generated by public phones is originated by people who do not own a telephone. Cave, Milne and Scanlan (1994 p. 11) however point out that while those who substitute call-boxes for home phones are mostly in low income groups, pay-phones are increasingly being designed exclusively for credit card holders. Additionally, the price of calls originated from public phones is significantly higher than the standard tariff applied to residential lines. The number of coin operated public phones may become a more appropriate measure of this social obligation in the future, and the evolution of pay-phone call charges ought to be monitored. Mercury's withdrawal from the pay-phone market in late 1994 suggests that pay-phone provision should remain part of USO.

Offering targeted subsidies to particular customers is designed to fulfil the prime social objectives of reducing inequality, promoting democracy and increasing opportunities. These subsidies are usually aimed at low income groups (e.g. Lifeline in the US or the Light User Scheme in the UK), the elderly and the disabled. There are 4 million deaf people in the UK today, and 10 million are hearing impaired. This group of users, which is bound to increase in size as UK population grows, does not yet have a '999' number to dial in emergency which will connect them to an operator equipped to respond to the hearing impaired. The design of targeted subsidies becomes of paramount importance to ensure that social equity is not neglected when the industry is forced to improve efficiency. Penetration rates among target groups of users should be the focus for monitoring progress in this policy area. These rates should be kept in mind when assessing the overall impact of efficiency-enhancing changes in the traditional structure of telecom tariffs, such as rebalancing of rental versus usage charges.

Calculating the costs of USO

We believe that supervision of USOs should be a leading regulatory task. The regulator should provide a definition of the different USOs and specify the quality standards that are expected in telecoms service at any level. Without a clear definition of quality, affordable universal service can be an academic concept. As liberalization promotes entry of more efficient firms, which should therefore be able to achieve USOs at lower costs than the incumbent, it is sensible to award USO on the grounds of efficiency rather than tradition. The task of realizing different goals of USOs should be allocated to the firm that is able to meet the specified levels of quality at lowest cost. USOs should be measurable, their achievement and costs should be periodically monitored, and they should be put out to tender. This will enable a clear and dynamic vision of the impact of competition upon industry costs, as well as a firmer starting point for future discussions about extending USOs to more advanced services.

Efficient delivery of USO requires its cost to be known. But costing the delivery of USO is not straightforward. Differences in costing methodologies can yield significant cost variations. Differences arise from the adoption of different hypotheses about people's behaviour, for example about how users who drop off the network will substitute for the telephone line they no longer have, and network cost allocation. The main variations in costs concern:

- The time horizon chosen: long term versus short term costs.
- The viewpoint employed: users who would join the network with USO versus users who would drop off without it.
- The way common or joint inputs are attributed to a given service: are common costs to be included or would they be there irrespective of USO customers?
- The way revenue flows are attributed: how to treat incoming calls and how to evaluate substitution calls.

Universal service is concerned with the long term gains of an integrated society, both in the sense of an increasingly sophisticated and widespread use of the telephone infrastructure and in the sense of a fuller participation in democracy. It is logical therefore, when choosing between short, medium and long run costs, to choose a long run time frame.

Further differences in viewpoint concern how 'uneconomic customers' are taken into account. One assumption is that such customers would never be connected by an efficient operator in a free market environment, and hence the cost of serving them – and the USO associated with

them – amounts to the full cost of connection, maintenance and operation of their lines. *Fully distributed costs* (FDCs), which include costs
directly caused by the relevant service and a share of the common costs
attributed to it, are an example of this type of costing. Calculations of
FDCs can vary widely depending on the assumptions made when
attributing common and joint costs, and on the rate of return assumed
for each service. Alternatively, one can look at the sums saved if unprofitable customers were disconnected, and calculate the so-called *avoidable
costs*.

These methodologies are likely to produce different estimates of the
costs of USO. But, however obtained, cost estimates must be related to
the revenues likely to be raised from customers targeted by USOs. The
actual bill paid by the customer, the value of incoming calls received by
the customer, and the value of the benefits to all customers on the network accruing from their ability to call the USO customer (i.e. positive
externalities) should all be factored in. There are many ways to estimate
the size of these three components. A customer who drops off the network may use other lines (work, friends or relatives) and this substitution effect must be taken into account. To estimate revenues fairly, total
revenues (outgoing and incoming calls) must be adjusted in respect of
replaced calls – both outgoing and incoming. The net revenue thus
obtained is called *avoided revenue*.

Cave, Milne and Scanlan (1994 p. 38) suggest a costing methodology to
compare avoidable costs with avoided revenue to obtain net USO costs.
The fact that the UK (and European) telephone network is already rolled
out, and most common costs have already been met before the relatively
small number of unprofitable USO customers is connected, suggests that
the avoided cost methodology is preferable to the FDC methodology.
However, a network which is under development is characterized by
much higher costs – economies of scale may not be fully exploited and
common costs are shared among a smaller number of customers – and
therefore increasing the number of customers would require costly
investment in the network without a proportional increase in revenues.
In such cases the FDC methodology may be preferable. Costs are likely
to change in time as network rollout takes place, and universal service
becomes less expensive and thus easier to achieve.

Providing service to the needy

In a study on elasticities of demand for access to the telephone network,
Cain and MacDonald (1991) show that many of the normal assumptions
on elasticities no longer hold for the 'borderline' class of customers who

might drop off the network. Flexible and innovative tariffing can transform 'needy' customers into viable telecom users, making a positive contribution to amortizing the costs of the network. Fod Barnes of Oftel has developed a neat worked example which shows both that flexible tariffing can benefit all users of a network and that, if a competitive telecommunications regime exists, regulation is required to inhibit cream skimming.

Barnes asks us to imagine a hypothetical two exchange network with each exchange serving six customers. In each group, two customers are heavy business users and four are light residential users. The total annual cost of the network, £1,920, is shared equally between the subscribers: that is, each pays £13.34 per month for service. If £13.34 per month is unaffordable for the eight residential subscribers, they will leave the network, and the total costs will be borne by the four remaining business subscribers. The costs of running the network may be lower with fewer telephone subscribers but, since most of the network infrastructure is still required, costs are unlikely to fall far enough to avoid an increase in service costs for the remaining four telephone subscribers. Let us say that the costs fall to £1,200 per year, that is, £25 per month per remaining subscriber. Barnes shows that in this case all benefit if the light users remain on the network and bear a share, even if an unequal share, of the total costs of the network. If residential users paid £7.51 per month and business users paid £24.98 all would benefit. Admittedly, in this extreme example the financial benefit to business subscribers is very small but they would also benefit by being able to call (and be called by) the residential subscribers. Innovative and unequal tariffing can benefit all users. This example demonstrates that universal service can yield economic benefits to all network users as well as benefits to society as a whole. For users who are likely to remain uneconomic USO funds can compensate the network provider for any losses incurred in providing service.

Where subsidies are required, targeted subsidies appear to be more efficient than generalized cross-subsidies, but may fail to provide an effective solution if they are not designed and publicized appropriately. A study undertaken by the Policy Studies Institute (PSI) (1995) reinforces this point: awareness of BT's Light User Scheme is very low in households without a telephone, and among those who had trouble paying their bills. Only 12 per cent of households which have never had a phone had heard about the scheme, compared to 30 per cent of households who had previously had a telephone. Despite being the second largest UK spender on advertising in 1994, BT failed to raise awareness of the Light User Scheme among the potential customers most likely to benefit from it.

The PSI study coupled a quantitative survey to identify households without a telephone with qualitative research to determine why potential users were untelephoned. It provides very useful groundwork for the design of targeted subsidies for Britain's low income households. PSI estimates UK telephone penetration to be around 93 per cent, close to Oftel's (1994b p. 4) estimate of 91 per cent which was based on 1993 data. Households without a telephone are poor households. The untelephoned are likely to be single people or lone mothers with dependent children, to live in local authority or housing association accommodation, to live in the North of England, to be young (16–24) and to be Afro-Caribbean.

Nearly half the 'untelephoned' formerly had a telephone, and either have been disconnected (17 per cent of those who no longer own a telephone) or have decided to do without a telephone service. The common view is that telephone ownership is not dynamic, but households constantly hop on and off the telephone network, mostly because of inability to pay the bills. However, in many instances such inability is not permanent. A change in the 'normal' conditions of the households, for example divorce or redundancy, may suddenly strain users' finances, particularly when users feel unable to control their debt to the telephone operator. Interviewees report the lack of any suggestion to help them maintain telephone services, such as to use call barring on premium numbers. PSI's qualitative interviews stress a sense of helplessness; PTOs do not offer alternative paying regimes and those who temporarily drop off the network face a reconnection charge of £50 and often a request for a £200 deposit for reconnection. Such practices are incompatible with the social goal of extending telephone penetration among needy users. Moreover, the high costs of connection to the network (in 1995 the BT charge was £99 plus VAT) inhibit extension of service to the untelephoned. Disconnection policy, and PTOs' handling of users who have difficulty paying, should be evaluated as an important part of the universal service obligation.

Figure 3 shows that difficulty in managing telephone bills is the most frequently mentioned reason for not having a telephone; it was also the most common reason for respondents giving it up. This finding brings out the most important point about the Light User Scheme: it is designed for the needs of people whose usage pattern, rather than spending ability, is low. BT has criticized the scheme as an 'inefficient way of increasing and sustaining penetration', recognizing that a 'variety of packages' and service options would yield better results (British Telecom 1995 p. 25). We share the belief that inventive marketing can increase penetration rates on a commercial basis and that, rather than imposing one particular tariff scheme, which may well be counter-productive, operators must be given the flexibility to market their services in a range of different

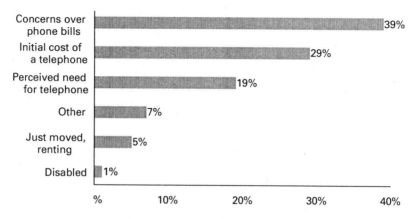

Figure 3   Main reasons for not owning a telephone
Source: Policy Studies Institute 1995

packages.[5] We also regret, however, that BT's efforts to keep 'unprofitable' customers on the network using the existing range of tools available to it are minimal. People who find it difficult to stay on the telephone network need a system to help them control the costs of a telephone, and limit the extent to which a change of situation (e.g. becoming unemployed) or an unprecedentedly high bill (the children calling premium services) may lead to disconnection.

The main obstacle to reduce disconnection is users' lack of information about their alternatives. Those who have difficulties in paying do not know whether they have a helpline and, if so, where to find it. Now that a collection of operators rather than a single one is accountable for all telecommunications services, it is all the more important that a well advertised consumer interest unit is set up to advise people with paying difficulties. Although it is in the commercial interest of all operators to market their entire range of tariffs, there are grounds to require the entire industry to set up a common helpline, via a *super partes* body, such as the regulator or the National Consumer Council. Increased awareness of the existing range of services and tariffs can only benefit competition, by improving consumer response to the competitive interplay of telephone companies, in a service area where consumer inertia can be a significant barrier to entry.

The regulator should set target penetration rates for particular classes of customers, and periodically monitor the extent to which USOs are achieved.

USO for broadband services?

The composition and extent of the USO will always be a matter of social and political judgement. We have discussed USOs in terms of voice telephony. However, other services must, in time, become part of the USO basket. As advanced telecommunications become increasingly important, more and more people will be able to access such services and those unable to do so will become more and more excluded from the mainstream. Democracy and equity demand that no citizen is unduly penalized through exclusion from such services. The EU has emphasized the need to minimize social tensions and risks associated with the transition to an increasingly wired society. But USO for broadband services *now* is a very expensive commitment and a poor answer to the rise of an information-rich/information-poor divide in society. Greater emphasis on IT education addresses this problem more effectively than universal availability of advanced transmission technology!

Although we recognize that the content of USO will be a matter of judgement, we believe that a mechanism to identify the services that should be part of the USO is required. Provision of new services to households should be left to the industry in the first instance. Market forces will indicate where technology is leading and which services people value. Only when services are widespread and commonly used does their availability become an issue on democracy and equity grounds. The extent to which services have been taken up by unsubsidized subscribers indicates which services should become part of the USO. If rapid take-up of new services is desired, public education programmes can stimulate rollout.

The Australian Bureau of Transport and Communications Economics (BTCE) (1994 pp. 93–8) proposed five steps for identifying whether a service should be included in the USO. We have refined BTCE's idea and propose that a service should be included in USO when these criteria are satisfied :

- Widespread use: the service should be accessible to 70 per cent of potential users and at least 50 per cent of users should have taken it up.
- The service displays characteristics of public goods or positive externalities.
- There is an absence of alternatives.

The regulator would evaluate a communications system according to these criteria, and decide whether USO can be implemented practically and efficiently. This framework and these criteria can be applied to telecommunications, broadcasting and other communications systems, as follows.

Widespread use   At present 98 per cent of households have a television (Office of Science and Technology 1995b p. 16), 6 per cent of British homes are connected to cable and 14 per cent are equipped for satellite television. As for telephones, 93 per cent of UK households have a telephone line and 75 per cent[6] of all exchange lines are digital. But only 3.5 per cent of UK homes have a modem, and are therefore able to access the Internet (*Computing* 24 August 1995).

Externalities or public good characteristics   Network externalities are important in telecommunications. In broadcasting, my enjoyment of a television programme is not directly affected by the number of people watching it, but there are positive externalities for people sharing sources of information and entertainment. The real economic reason for universal service provision is that broadcasting is a classic public good: the marginal cost of an extra viewer is nil, and the more people watch television, the cheaper it is to deliver a particular programme to a particular viewer.

Absence of alternatives   This is true for both basic and enhanced voice telephony: there is no valid substitute for the ability to manage costs provided by itemized bills and call barring. In radio and television, the diverse and impartial generalist channels delivered by terrestrial broadcasting are today the object of USO. Alternative delivery systems do not offer channels characterized by the same content requirements and in this sense they are not a substitute for terrestrial broadcasting. Perhaps plurality and diversity of content in broadcasting will eventually be satisfied by a plethora of dedicated or even interactive channels rather than by a bouquet of generalist ones, as they are today. Digital broadcasting forces us to look at USO in this new way, making choices between types of broadcasting and selecting for USO which responds best to the needs of entertaining and informing the public.

Overall, terrestrial broadcasting qualifies as an essential service, more widely used though perhaps less useful than telephony. The Internet and other (cable based) interactive services do not qualify: they are not widespread enough.

Summary

1 USOs are rules imposed on suppliers of basic services and utilities in order to ensure that the service is available to all at affordable cost.
2 Questions of USO are urgent in telecoms, because the incumbent firm charged with them is no longer a monopoly. Efficiency and cost of universal provision must be spelled out and monitored. In broadcasting, the USOs are challenged by the advent of digital technology and the expansion of alternative delivery systems.
3 USO in broadcasting is one aspect of the more general entitlement to a right to cultural identity. In the UK this right is obtained by two public sector operators (the BBC and Channel 4) and by positive content requirements imposed on commercial operators in exchange for spectrum. All terrestrial broadcasting services are also universally accessible and free at the point of use.
4 The sustainability of these arrangements is likely to decline in the near future. The licence fee will be called into question, as the supply of TV services increases and the BBC's share of viewing declines. The attractiveness of spectrum for advertising funded broadcasters is also likely to decline as alternative delivery methods reach greater audiences.
5 Hence we propose maintaining publicly funded broadcasting services and imposing 'must carry' rules on other terrestrial services that fulfil positive content requirements.
6 USO in telecommunications is usually taken to mean an obligation to widespread network rollout. USOs also include special services (emergency calls, services to disabled), public phones and special subsidies to light users.
7 These different policies should be individually spelled out and costed, then imposed on the operators who can meet those costs most efficiently. The funding should come from the entire industry.
8 Special subsidies should be narrowly targeted at the receiving group, to be effective. The Light User Scheme, for example, helps those who use the phone very little rather than those who need the phone but cannot pay for it.
9 The content of USO should change as new services become widespread. We suggest a mechanism to select new services that should be candidates for future inclusion. At present, none of the broadband services would pass our test.

5 Freedom of Expression

Regulation is mainly concerned with monitoring and correcting the action of firms and markets. Only rarely are regulators interested in the content of the products in the regulated markets, beyond concerns for health and safety. In this respect, media are exceptional. Whether for reasons of security, or taste and decency, or because of spectrum scarcity, regulators have often posed limits on the content of media products. This chapter will explore how far restrictions to freedom of expression can be justified. The aim is to identify general principles that can be applied to varying circumstances. Current regulatory restrictions presuppose a distinction between the private and the public which is increasingly difficult to draw. Current restrictions may rest on technological conditions that are changing. Too often, they represent the views of unelected gatekeepers who have the power to decide for all but are accountable to few.

Freedom of expression in the United Kingdom[1] is restricted by at least a dozen Acts, but is nowhere guaranteed. Our freedom of expression is curtailed by numerous specific statutes and by common law provisions – notably in respect of the laws of defamation and confidentiality (Lord Chancellor's Department and the Scottish Office 1993 pp. 13–19):

- Obscene Publications Act 1959 and 1964
- Indecent Displays Act 1981
- Contempt of Court Act 1981 (revelation of sources)
- Police and Criminal Evidence Act 1984 (revelation of sources)
- Video Recordings Act 1984 (video classification)
- Race Relations Act 1965, then Race Relations Act 1976, now Public Order Act 1986 (incitement to racial hatred)
- Malicious Communications Act 1988 (material distributed by mail)
- Official Secrets Act 1989

- Prevention of Terrorism Act (Temporary Provisions) 1989 (revelation of sources)
- Broadcasting Act 1990 (notably in Section 10)
- Football (Offences) Act 1991 (indecent or racist chants)
- Criminal Justice and Public Order Act 1994 (right to assemble, video classification).

If freedom of expression is not absolute, a power is given to the person who decides where the line is to be drawn. If that person is an agent of the state, careering down the slippery slope of censorship may be irresistible. In the UK, we have slid too far. In too many cases, the interests of the state or its agents are allowed to override the freedom of expression of individuals, both in public and in private communication. This harms our fundamental freedoms and protects the state from justified scrutiny.

To avoid this, freedom of expression and access to official information should be entrenched as rights. This would put the onus on the state to demonstrate the public interest when overriding those rights. In this chapter we propose a right to freedom of expression and to government information and examine the implications of these proposals for British law. We start from the following principles:

- Government should not be able to interfere with the right of individuals to hold opinions, or to communicate them privately.
- Individuals should have the right to express opinions publicly, subject to certain suppressions and restrictions necessary to prevent harm to others.
- Prior restraint should only be permitted where there is a clear and present danger – i.e. when there is a severe threat to national security or public order.

Many potential issues are involved in freedom of expression, from censorship of the press to the right of assembly and access to government expression. We will focus on the media. Issues of individual freedom of expression and access to information will be considered only as they are relevant to the media.

Behind debates on freedom of expression, there are two main types of argument: those based on rights and those based on consequences. When Andrea Dworkin (1993) says that pornography is the theory of which rape is the practice she is appealing to consequentialism. When the Illinois Supreme Court ruled that the Nazi Party could parade through the Jewish neighbourhood of Skokie, it put the right to freedom of expression before possible consequences.

However, these theoretical perspectives are not so neatly opposed. Consequentialism can be used to defend freedom of expression, for example where Erica Jong (1996) argues that without freedom of expression there is no great art. Rights theories can be used to defend censorship, for example where Catherine Itzin argues that pornography should be controlled because 'the right of women to be free of the misrepresentation and mistreatment of pornography is, I think, a fundamental human right' (1994 p. 20).

In practice, most people refer to both theories simultaneously. Even rights absolutists in the US Supreme Court have allowed exceptions 'where "fighting words" are concerned ... or where privacy is at stake in defamation law, or where commercial competition is in issue and where misleading speech could adversely affect the market' (Lee 1990 p. 40).

Rights and consequences should both be taken into account. The logic of combining the two theories is best seen by looking at the dangers of adopting one exclusively. A rights absolutist could not override the right to freedom of expression, however certain and harmful the consequences might be. The absolute consequentialist, on the other hand, would need to find a method for predicting consequences and deciding which ones should not be tolerated. This could not be done democratically because the decisions of the majority could override the basic interests of an individual. Nor could it be done by the executive, because politicians will be tempted to use censorship to prevent scrutiny of their activities.

We should have a right of freedom of expression. But what would that right mean? A right protects the individual against mob or state power, but can be ignored in the public interest in exceptional circumstances, as IPPR made clear in arguing for a right to freedom of expression as part of a constitution for the United Kingdom: 'Traditionally in the English legal system, since the right to freedom of expression has no enforceable constitutional protection, the protection of the interests of the state or its agents can override the freedom of the individual . . . [A rights based] approach is quite the reverse, guaranteeing the individual the right to freedom of expression unless a restriction on it can be shown to be justified' (1991 p. 11).

Customarily we claim rights when we perceive inequity. If I am confident of my position in society, I am more likely to tolerate speech criticizing me. If I am part of a minority that I feel is discriminated against, or excluded, I am likely to be more sensitive. For example, the British Muslim community could not invoke the blasphemy laws against *The Satanic Verses* (Rushdie 1992), because the law only recognizes anti-Christian blasphemy. In the absence of equality of treatment, mainstream Muslims are said to have found it hard to hold a tolerant line against pressure from fundamentalists. The real question at issue was as much inequality between communities as the right to freedom of expression. By resolving the fundamental problem – sexual or religious discrimination for example – we can create a culture where freedom of expression is perceived to be less threatening. However, we will never be able to accommodate all minorities. In a mixed society like the UK, there will always be fundamentalists who object to standards which the rest of society finds tolerable. While we should try to protect minorities, we cannot impose on society the standards of the most easily offended. How can we help those communities without insinuating intolerance into the law and giving both the government and the fundamentalist the right to censor? Who decides? The answers to these questions are discussed below.

A right to freedom of expression

The right to freedom of expression should be entrenched in a bill of rights. The point of a right is to put a demanding requirement on the state to prove the public interest when seeking to override it. If that right is merely given in legislation, it could be easily overridden, or be deemed not applicable to other Acts. Putting a right in a constitution makes that right primary to legislation. It makes the right 'open textured', i.e. capable of guiding law makers and the judiciary when coping with unforeseen circumstances or conflicting objectives (IPPR 1991 p. 13).

Two main examples of rights to freedom of expression are the European Convention on Human Rights (ECHR) and the UN's International Covenant on Civil and Political Rights (ICCPR). However, the formulation of the right to freedom of expression in the ECHR allows wide-ranging consequentialist exceptions:

> The exercise of [the right to freedom of expression], since it carries with it duties and responsibilities, may be subject to such formalities, conditions and restrictions or penalties as are prescribed by law and are necessary in a democratic society, in the interests of national security, territorial integrity or public safety, for the protection of disorder or crime, for the protection of health and morals, for the protection of the reputation or rights of others, for preventing the disclosure of expression received in confidence, or for maintaining the authority and impartiality of the judiciary. (Article 10.2)

This article does make clear that any derogation must be 'necessary' and provided for in law. But the let-out clauses are not just numerous, they also define very broad grounds on which governments can decide to censor opinions – particularly the clause allowing for the 'protection of health and morals'.

The UN's International Covenant on Civil and Political Rights comes closer to providing a substantive right. Unlike the ECHR, it specifies that 'everyone shall have the right to hold opinions without interference'. This is a vital distinction between the right to hold an opinion, which cannot be restricted, and the right to express that opinion, which can. We take the view that non-criminal exceptions should be made possible, but without giving the government an implicit right to judge and enforce morality, to decide which opinions can and cannot be expressed. Instead, expression should be limited only where it will cause, or may cause, harm. Moreover, although the IPPR model uses a superior formulation to that in the ECHR, pragmatic considerations might suggest a simple adoption of the ECHR into UK law. Such a measure would improve protection of rights in the UK; would remove a significant discrepancy between UK and European law; and would enable the UK to benefit from the case law that has accumulated around the ECHR.

Since there is no ideal formulation of the right to freedom of expression available 'off the shelf', we need to create our own. To do so, we must be clear about the principles we want it to embody and the exceptions we need to allow. This section suggests the principles; the rest of the chapter examines the exceptions that need to be made.

In 1977, the then Labour government set about a similar task. It asked Bernard Williams to chair a Home Office Committee on Obscenity and Film Censorship (Home Office 1979). By the time the Committee reported, there had been a change of government: the audience was less receptive and few of the report's recommendations have been implemented. Yet those recommendations have stood the tests of time, changing technology and attitudes. They could provide a firm basis for a new constitutional settlement on freedom of expression.

Williams described the principles formulated by his committee:

> The printed word would be neither banned nor restricted for reasons concerned with pornography. Secondly, a small class of works could actually be suppressed ... where there was a presumption that a crime was committed in the course of making it ... [Thirdly,] matter should be restricted which not consisting of the written word, is such that its unrestricted availability is offensive to reasonable people by reason of the manner in which it portrays, deals with or relates to violence, cruelty or horror, or sexual, faecal or urinary functions or genital organs. (1996 pp. 25–6)

Two building blocks of the Committee's principles are noticeable. First, Williams talks about suppression and restriction instead of censorship, which can mean either. We will use suppression to mean prohibiting the ownership or distribution of certain material. Restriction is taken to mean the imposing of conditions on the distribution of the material – for example on its location, on the age range of its consumers, on the time at which it is shown, or on the punishment that may be appropriate if distribution is shown to have caused harm. Second, the definition of offensiveness turns on the tastes of 'reasonable people'. This does not solve the problems of coping with offensiveness in a multi-cultural society, but it does make clear that the standard is not that of the most easily offended.

The Williams Committee's framework presumes in favour of freedom of expression, only allowing suppression where harm can be demonstrated. On the other hand, it recognizes the legitimate interest people have in not being offended and provides restriction as a light touch solution to those concerns. This framework seems to go with the warp of public opinion: a survey by the regulator of telephone premium line services ICSTIS[2] found that the public objected more to the adverts for chat lines in 'family newspapers' than they did to the contents of those products, which were predominantly confined to a volunteer public (1993 p. 5). The basic principle underlying the attitudes of both the public and the Williams Committee is to restrict material that causes offence, but not to suppress it without clear evidence of harm being done.

But the Williams approach was targeted at pornography, and is therefore not the complete answer. How could this framework be applied to race, religion, national security, privacy, terrorism? And does new technology invalidate any of the principles laid out in 1979? In examining these areas, we must not forget that legislation is only one side of the coin. Much can be done for freedom of expression by remedying inequalities and encouraging a tolerant culture.

Restrictions to freedom of expression

How grave do the effects of freedom of expression have to be before we take them into account? Mary Whitehouse and her successors at the National Viewers and Listeners Association are fond of quoting 200 studies which show that television causes violence, abuse and denigration. From a very different ideological starting point, Catherine Itzin (1994) argues that there is a correlation between pornography and harm to women and children, though it is impossible to *prove* that the consumption of pornography causes harm.[3]

To say that freedom of expression cannot be curbed, whatever the consequences, is mistaken. If a film could be demonstrated to be causing copycat murders across the country, there would be grounds for its suppression. Our preferred option is to stress the importance of a right to freedom of expression, while recognizing that it can be limited where identifiable harm is caused. But what kind of effects justify suppression?

There is a narrow range of cases where the likely harm is sufficient for speech to be suppressed before publication. This is when the danger is 'clear and present' and cannot be remedied through discussion after publication. National security and public order provide the obvious instances where suppression may be justified. However, case by case judgement is needed and specific decisions are always likely to be controversial.

More controversial is whether television violence or pornography could fall into the category meriting suppression. It is not immediately clear that the studies Mary Whitehouse adduced provide evidence of a clear and present danger to others. Some commentators argue that research does not demonstrate a clear link between television and violence.[4] They challenge the methodology of the studies claiming such a link.

In fact, it is at least arguable that liberal attitudes to pornography and violence are correlated with *reduced* levels of harm. For example, a recent study found no consistent relation between high levels of television violence and high levels of actual violence. Osmo Wiio (1995) compared murder rates with television ownership, viewing and proportion of violent programming, across forty-five countries, and found no consistent correlation between murder rates and television viewing, or between murder rates and the proportion of violent programmes on television. Indeed, he found a negative relationship between violence and television ownership. The only country with a positive correlation is the USA. Wiio contrasted the USA with the UK, which has the second lowest murder rate in the countries studied, but high levels of television ownership, viewing and violent programmes.[5]

Hence, arguments for censorship remain unproved, and no one has demonstrated that pornography or violence in the media *in general* creates a clear and present danger to others. Until they do, blanket suppression of types of material cannot be justified. That does not undermine Itzin's argument for prosecution of particular magazines or films, though harm would require to be demonstrated rather than presumed.

In summary, entire categories of material should not be restricted or suppressed. Exceptions to freedom of expression can be made in individual cases where the right to freedom of expression is outweighed by the identifiable harm it causes:

- To suppress before publication (i.e. impose a prior restraint), it must be shown that the utterance presents a clear and present danger that cannot retrospectively be remedied.

- To suppress after publication, it must be shown that a harm has been committed.
- To restrict minors' access to material actual harm need not be shown, but the protection of minors should not deny adults access to material to which they otherwise have lawful access.
- To restrict material on behalf of adults, it must be shown that reasonable people are likely to be offended by the material.

The next sections apply these principles to expression that may harm the country's security, racial, religious, or gender minorities, and individuals' privacy.

National security

There are uncontroversial examples where speech causes a clear and present danger to national security – i.e. the first of our four tests. Revelations about military secrets or about counter-intelligence work could directly endanger lives and would be irrecoverable after disclosure. The right to freedom of expression may, therefore, legitimately be limited to allow prior restraint on the grounds of national security.

However, national security should not be confused with the government's security, particularly by the government of the day. There are numerous examples in the UK where national security has been invoked in controversial circumstances.[6] In particular what concerns us here is the law which allowed the government's actions.

The Official Secrets Act 1911, did not allow a public interest defence and nor does its 1989 successor. The 1911 Act treated all government information the same: none of it could be revealed. The 1989 Act at least has the merit of recognizing that there are different levels of official information. Unfortunately, virtually everything is placed in the category that should be kept secret. Government records are inaccessible for 30 years, and the Lord Chancellor can decide to extend that period to 50 or 100 years or indefinitely. There is no appeal against his decision. In 1993, the government proposed a new code of practice on access to government information, which gives the public access to some types of non-classified material and gives MPs a right of appeal to the Parliamentary Ombudsman about breaches of the code (Chancellor of the Duchy of Lancaster 1993). Liberty argues that the government is still too powerful: too much information can be restricted for too long too easily. In particular, Liberty points out that central government has imposed stringent openness requirements on local government, but not on itself (Local

Government (Access to Information) Act 1985 and Local Government and Housing Act 1989, as quoted in Foley 1995 pp. 291–2).

A right of access to official information, which could then be balanced against necessary exceptions, would be a better remedy. We propose a right to freedom of information below.

Terrorism

If terrorist threats are threats to national security, restrictions on freedom of expression in the name of national security may be justified when the release of information would cause a clear and present danger. However, terrorism does raise separate censorship issues. UK governments of both parties have covertly and overtly suppressed programmes which posed no clear and present danger (see Smith 1972; Schlesinger 1987). Is terrorism a special case?

One argument says that terrorism is more than a threat to national security: it is an internal war. Margaret Thatcher used this argument to justify the 1988 ban on broadcasting the speech of representatives of Sinn Fein and other terrorist organizations when she said that: 'In order to beat off your enemy in war, you have to suspend your civil liberties for a time' (*The Times* 26 October 1988).

However, the dangers of this approach are easy to see: the South African government used the same argument to justify its reporting restrictions during the township troubles of the late 1980s. Terrorism is a special case, because the point of terrorist outrages is to create publicity for a cause. The reporting of such 'armed propaganda' may therefore contribute to its incidence. However, such consequences do not justify the type of power the government used to impose the broadcasting ban. Admittedly, the broadcasting ban did not suppress, it restricted. But in peacetime, the government should not be able to impose such a ban by notice, as it did. Under the terms of the BBC Licence and the Broadcasting Act 1990, a notice is a ministerial order that broadcasters must obey and which the government can use to stop the showing of a programme on any ground it chooses. It is subject to challenge in the courts but only on the ground that the Minister has acted unreasonably or perversely. As Barendt (1994 p. 27) has pointed out, such a provision would be unconstitutional in both Germany and the US, and no other part of the media is subject to such draconian censorship. The fact that the power has been little used does not undermine the case for its repeal.

Racial hatred

Apart from national security, the most unambiguous exception to the international law on freedom of expression is incitement to racial hatred. In fact, the US only ratified the UN's International Covenant on Civil and Political Rights recently, because, before Bill Clinton became President, these exceptions were thought to be incompatible with the First Amendment to the Constitution of the United States of America: 'Congress shall make no law ... abridging freedom of the speech or the press.'

Do laws which penalize incitement to racial hatred unjustifiably restrict freedom of expression? Do we penalize incitement to racial hatred because it causes greater harm than other types of hate speech? Because there is clear evidence of a causal link between speech and harm? Or is speech that undermines the fundamental values of our society not worth defending?

On one level, the main issue is racial discrimination and violence. The Home Office estimates that there are between 130,000 and 140,000 incidents of racial attack and harassment each year (*Guardian* 15 July 1993). Racist attitudes exist in the United Kingdom, although within the political mainstream rarely overtly. Even if there are many causes of racist attack and discrimination, few would deny that racist attitudes contribute to racist violence. But the issue is not whether racist discrimination and violence should be allowed, but whether racist speech should be allowed. In turn, this question breaks down into three sub-questions:

- whether we want to prevent racist speech itself, or the harm it putatively causes
- what level of harm would justify limiting freedom of expression
- what types of limits are needed, and particularly whether and when racist speech should be suppressed.

Article 20 of the ICCPR provides one answer. It prohibits 'the advocacy of national, racial or religious hatred that constitutes incitement to discrimination, hostility or violence'. Article 20 restricts racists' speech independently of whether harm arises from it. New Zealand's Race Relations Act has a similar phrasing. This section examines whether our law should be extended in this direction.

We cannot protect any group from expression of hostility without severely qualifying freedom of expression. While discrimination and violence are both types of harm suffered by the victim, hostility is an opinion held and expressed by an individual. Article 20's prohibition of incitement to hostility conflicts with fundamental principles of freedom of expression and with the Williams position: that nothing should be censored by virtue of the opinion it expresses. It runs counter to the guiding principles of our policy.

Can an exception be justified? Prohibiting incitement to racial hostility would have two advantages. It would give a clear signal that racism is unacceptable in the British polity and it would reduce the feelings of victimization of many members of victimized groups. However, the phrasing of Article 20 would have clear disadvantages. It would establish a principle that governments can determine what opinions citizens can and cannot express. It would chill freedom of expression and allow censorship of, for example, radical black power movements. The benefits do not outweigh the disadvantages sufficiently to breach the principle that government should not suppress material by virtue of the opinion it expresses.

The second question is what level of harm speech has to cause before it is justifiable to curb it. If we believe ' hostility' is not enough, Article 20 offers us two further options: violence and discrimination. Violence and racial discrimination are punishable under civil and criminal law and the Race Relations Act 1976. What is at issue is whether a separate law to prevent incitement to hatred is needed, over and above those which criminalize violence and racial discrimination.

Harm resulting from incitement to hatred is not easy to prosecute, because not all incitement is immediate and individualized. Incitement to racial violence is targeted at a group of people rather than at an individual. It may have effects over time rather than at the moment of delivery. That makes the link between the incitement and the offence harder to trace, and therefore the crime harder to prosecute using just the law on violence and discrimination. The case for such a law, which is supported from a civil rights perspective by Liberty, therefore turns on the effects it prevents. Our approach is that prior restraint is justified only if its effects could not be remedied, in other words immediate violence and breaches of public order. UK laws on incitement to racial hatred are narrowly aimed at public order. The Public Order Act 1986 permits suppression of material which has been found to breach its provisions, even if received only by members of an association, and allows arrest of people who would use words or publish material that is threatening and abusive and intends to stir up racial hatred but does not allow prior restraint unless expression of racist sentiment is to be effected through a street demonstration or procession (Public Order Act 1986, Sections 10–11).

Indeed, there is concern that it is unnecessarily difficult to achieve a conviction under the current law. The Home Affairs Select Committee reported that only eighteen cases had been brought under Part III of the

Public Order Act between 1986 and 1994. They accepted the argument that this reflected the difficulty of using the law, rather than a low level of offences committed under it.

A new offence of group defamation?

The Tabachnik Report from the Board of Deputies of British Jews (1992) argues for the creation of a new offence of group defamation. The Tabachnik Report makes a persuasive case that this change would bring us more closely into line with other countries' laws. Tabachnik argued that, in its exclusive focus on public order, the law of incitement in Britain differs from that in a number of other countries (pp. 10–20). In Canada, incitement to hatred is prohibited in itself, but very wide defences are available. In Belgium and France, properly constituted groups are allowed to bring proceedings for incitement to hatred. French law contains an offence of group defamation.

An offence of group defamation would not contradict fundamental principles of British law, according to the Board of Deputies. They argue that it seems illogical that material that is proven to have defamed an individual can be repeated with impunity about a group. Why is it unacceptable to say that 'X, an Englishman, is stupid', but acceptable to say that 'All Englishmen are stupid'? The Tabachnik Report calls for a criminal offence of group defamation to be created and argues that it would have the following advantages:

- remedy of harm to racial groups' reputations
- deterrence of future harm
- clear standards about what is acceptable in a democratic society
- no necessity for prior restraint
- no conviction without proof of harm (to reputation of the group).

However, there are also clear and substantial disadvantages to the Tabachnik proposals. The reference to group reputation could open the door to a wide range of material, not all of it harmful, wilful or untrue. The law could allow frivolous prosecutions. It might not gain widespread support and might therefore act as a focus for racist feelings.

If the case for such a law is accepted by government, it should be preceded by a consultation paper and full public debate. This would test public support and allow the legal aspects of the proposal to be refined. As the Tabachnik Report says: 'we are not committed to any particular form of words so long as any formulation embodies the principle that the law should protect minorities against group defamation' (p. 60).

On balance we agree with Brittan that 'It is neither desirable nor possible for the law to provide a remedy for all the unkindness of man to man' (Brittan 1963, cited in Lord Chancellor's Department and the Scottish Office 1993 p. 21): the potential danger to freedom of expression entailed by a new law of group defamation, which would punish general abuse rather than specific defamation and injury, outweighs the benefits likely to accrue from creation of such an offence. Voluntary restrictions could be more effective than statutory ones. These are becoming more widespread, as newspapers realize that they have a commercial interest in not offending their ethnic minority readership. Moreover, there are other ways for society to demonstrate its rejection of racist and other offensive behaviour, for example, through advocacy of tolerance and racial equality in public life and through the formal education system.

Blasphemy and religious hatred

British law does not protect religious groups or nationalities. But it has protected Christian belief since 1676 through the application of blasphemy law. The law protects only Christianity. In Scotland, no prosecutions have occurred since the 1840s and in England only one case (in 1978) has been prosecuted since 1922.

It should be uncontroversial that a law against blasphemy protects non-Christian religions. In the UK church membership of non-Christian religions exceeds that of the Church of England (Central Statistical Office 1993 p. 153) and, on current demographic and membership trends, Islam may overtake the Church of England within two or three decades (Modood 1994 p. 58). The exclusion of other religions from the blasphemy law only fans the flames of fundamentalist arguments that religious minorities are not bound by English law – for example, when putting a *fatwa* above the criminal law. Two Church of England working parties have agreed that the law should be extended to all religions. One in 1981 argued for the extension of the blasphemy law. The other in 1988 argued for a new offence of 'publish[ing] grossly abusive or insulting material relating to a religion with the purpose of outraging religious feelings' that would apply to all religions.[7] The inclusion in Article 20 of ICCPR of national, religious and racial groups has the advantage of theoretical consistency. Why then include racial groups but not religious groups? And why is incitement to religious hatred prohibited in Northern Ireland, but not in England and Wales?

There are various arguments against an extension of the law in this direction. First, the principle of protection of religious groups is implicit in current British legal practice, for example in the *Mandla* v. *Dowell Lee* case ([1983] I All ER 1062) which found that Sikhs are protected from racial discrimination, even though they are arguably a religious grouping. Second, the parallel with Northern Ireland does not hold: Northern Ireland has 75 per cent church membership and a recent history of sectarian strife and violence, whereas England has an 11 per cent church membership rate and no prominent religious violence or disorder of its own (Modood 1994 p. 58). Finally, the Tabachnik Report argued that it would be difficult to define a religious group and that such an effort would be superfluous since religious groups such as Sikhs and, the report argues, Jews are already covered by racial hatred legislation following *Mandla* v. *Dowell Lee*. The Law Commission agreed that without a clear need for this extension, the law should remain unchanged.

The real question is whether we should have a blasphemy law at all, not whether it should be extended. Consistency of treatment between religions should be achieved by abolishing protection for Christianity, rather than by extending the protection enjoyed by Christianity to other beliefs. No religion should be given prime place: the blasphemy laws should be repealed, not extended.

Pornography

We define pornography as displays of explicit, sexual nudity or activity, where the display is the end in itself rather than a means to a different end (for example artistic or educational). Some commentators argue that pornography incites and causes sexual violence. The UN Committee on the Elimination of All Forms of Discrimination Against Women has argued that 'pornography and the depiction and other commercial exploitation of women as sexual objects ... contributes to gender-based violence'.[8]

On the other hand, a recent Home Office commissioned review of the research on the effects of pornography found no evidence to suggest that eliminating porn would result in the decrease of sexual attacks against women and children. Though there is evidence 'that sexual offenders tend to be more involved with extreme forms of pornography' (Howitt and Cumberbatch 1990 p. 83), it is far from clear that pornography causes the crime. Which way does the chain of causation run? Do those predisposed to sexual offence seek out pornography? Or does consumption of pornography lead to sexual offences? Given the absence of firm evidence that consumption leads harm, the case for suppressing pornography is not robust.

There is a qualitative difference between pornography and incitement to sexual violence. Although some pornographic material may incite violence, not all pornography does so. A number of anti-censorship feminists have argued that pornography can liberate as well as upset and discriminate. As Erica Jong reminded us: '"Pornography", says Susie Bright, "is necessary to liberation. Pornography", says Sallie Tisdale, "is desired by women as well as men"' (1996 p. 15).

Not all pornography is harmful, so not all pornography should be banned. Those who want to ban certain types of pornography need to make very clear the reasons why we should support such a ban.

In certain circumstances, there are grounds to believe that the production of pornography causes harm, for example when the pornographic material depicts a criminal act. For this reason Williams suggested suppressing child pornography and pictures of actual rapes. In such cases, we can legitimately say that pornography caused harm, which would not have happened but for the production of the pornographic material. Equally, suppressing the material, by reducing the commercial incentive to photograph such acts, might reduce their incidence. Since any harm would already have been committed, prior restraint would not be justified.

Williams maintains that the burden of proof should be on the producers of such material.[9] The gravity of the offence and the fact that it is far easier for the defence to show that the participants consented than it is for the prosecution to show the reverse, justify departing from the normal assumption of 'innocent until proven guilty'.

Explicit and intentional incitement to sexual violence is different from pornography. If such incitements exist, extension of the incitement law to include victims of sexual violence may be warranted. With such extension, expression which may cause harm could be retrospectively punishable, but not suppressed.

Non-violent pornographic material, however, may still offend some. For example, most people accept that parents should be able to protect their children from porn or even non-pornographic material that contains nudity or depictions of sexual activity. Hence, pornographic material should not be accessible to children, and material that might offend should be restricted to a volunteer audience. The Williams Committee found that 'there was a very broad consensus that the main objective of the law should be to protect members of the public from the nuisance of offensive material in places to which normal life happens to take them' (Home Office 1979). We support present arrangements which confine the display of erotic material (soft porn) to top shelves, and sale of hard core material to specialized adult shops.

Restricting access to offensive material is best implemented through codes of practice drawn up by the regulator in partnership with consumers and producers. Those codes of practice should aim to give parents the means they need to control their children's access to porno-graphic material. Codes of practice will be more adaptable to techno-logical and cultural change than primary legislation: for example, the ITC has been able to decide, without recourse to Parliament, to exclude video-on-demand from watershed requirements on the grounds that it is restricted to a volunteer audience.

However, one question remains unsolved, particularly in the light of changing technology. Which media offer unrestricted availability? In her campaign against images of 'page three' girls, Clare Short MP did not argue that pictures of nude women should be banned. She argued that newspapers should be treated as public displays, and therefore be included in the scope of the Indecent Displays Act (Short 1996). The regulator needs to decide whether the present restrictions need to be extended so that such material is restricted to 'volunteer shops'. Judgement will turn on what reasonable people find offensive and what is meant by unrestricted availability. A new form of public involvement such as a citizens' jury may be an important part of the decision making process.

In any case, where material is restricted to a volunteer audience, the tastes of that audience should not be censored by the tastes of others. Here, the distinction between soft and hard core pornography (hard pornography shows an erect penis) should be abolished. This distinction tends to do more harm than good. By promoting pornography which concentrates on female nudity, this distinction tends to reinforce stereo-typed representations of women as passive objects. It also aggravates the bias against images of male arousal and full, mutual, heterosexual or homosexual activity.

Here, too, voluntary restrictions and responsible use of freedom of expression are preferable to government action. The Campaign against Pornography and Censorship recognize this when they encourage newsagents not to stock certain material but argue against suppression. Under the Local Government Act 1982 local authorities are empowered to ban or restrict sex shops and cinemas in their areas. Newsagents should be free to restrict offensive material. Any allegations that distrib-utors force retailers to stock pornographic material should be investi-gated by the competition authority.

Freedom of information

When discussing restrictions to freedom of expression which are justified by national security, we argued for a right to access public information. IPPR's model UK Constitution includes such a right (1991 p. 22). It allows only exceptions that are prescribed by law and necessary in a democratic society, namely those required:

- for the protection of national security
- in the interests of law enforcement or the prevention and detection of crime
- for the protection of personal privacy, legal privilege or commercial processes or transactions
- to enable a public service to perform its constitutional functions, or a public authority when acting in the capacity of regulator, contractor or employer to perform its functions.

A right of access to official information is an essential partner to a right of freedom of expression. Any derogation from the rights of freedom of expression or access to information on grounds of national security should be subject to a right of appeal and a public interest defence, whether through a parliamentary committee or through the courts. The appeal body should apply stringent standards: for example, the US Supreme Court has ruled that 'prior restraint of official secrets was not an option it would allow the US government unless lives would be lost' (Lee 1990 p. 49). Since the whole point of the derogation is to keep information secret, this proposed appeal procedure cannot be perfect, but seems the only way of balancing clear and present danger with open government. What is required is a UK Freedom of Information Act comparable to those in other European Union states, such as Denmark, France, Greece, The Netherlands and Sweden, and in other English speaking societies such as Australia, Canada, New Zealand and the United States.

We propose:

- the enactment of a Freedom of Expression Act including the provisions of the Right to Know Bill tabled by Mark Fisher MP in 1993 which would provide for access to official documents and expression (subject to legitimate exemption)
- appeal to an Ombudsman or Tribunal if access is refused (or unduly delayed)
- a public interest defence in cases of breach of the Official Secrets Act and the requirement that the prosecution shows that harm is likely to result in consequence of breach of the Act.

Privacy

Striking the right balance between protecting legitimate privacy and opening up what is properly public has long been a vexed question. The absence of legal protection of privacy in English and Scots law has provoked several parliamentary initiatives designed to protect individuals from unwarranted intrusion. Recently, the Lord Chancellor's Department and the Scottish Office (1993)[10] reviewed the issues and invited comment on the desirability of a civil remedy for infringement of privacy. The Lord Chancellor's initiative followed Sir David Calcutt's (1993 paras 7.37 and 7.38) judgment that press self-regulation did not work sufficiently well to justify the continuing absence of a tort of infringement of privacy.

Thus far, the UK has made no provision for legal remedy for infringement of privacy, although the Lord Chancellor's Department showed persuasively that existing legal remedies are insufficient to protect individuals' privacy. Significant changes have been made to the press self-regulatory agency, the Press Complaints Commission (PCC), though both the PCC and most of the press are opposed to statutory protection of privacy. Press conduct has not allayed our concern.

Privacy has been variously defined. The Lord Chancellor's Department's paper cited the *Oxford Dictionary* definition of privacy as 'the state of being private and undisturbed ... freedom from intrusion or public attention' (p. 8). The definitions based on a draft by Justice, and used in the Parliamentary Bills tabled by Brian Walden and William Cash, were more inclusive and more specifically tailored to media policy. The Cash Bill proposed a right of privacy described as 'the right of any person to be protected from intrusion upon himself, his home, his family, his relationships and communications with others, his property and his business affairs' (pp. 23–4). It defined intrusion as:

- spying, prying, watching or besetting
- the unauthorized overhearing or recording of spoken words
- the unauthorized making of visual images
- the unauthorized reading or copying of documents
- the unauthorized use or disclosure of confidential expression, or of facts (including his or her name, identity or likeness) calculated to cause a person distress, annoyance or embarrassment, or to place him or her in a false light
- the unauthorized appropriation of his or her name, identity or likeness for another's gain (pp. 23–4).

Clearly, even among those who believe that the public interest would be served by a legal remedy for infringement of privacy, definitions of the boundary between the private and the public differ in important respects. There is room for disagreement as to whether flexible definitions of privacy, such as that canvassed by Calcutt, whereby 'personal expression' is that which may legitimately remain private,[11] or tightly specified instances of intrusion, such as those specified in the Cash Bill, are to be preferred. Flexible definitions are able to accommodate change, whereas specific definitions give clear guidelines and thus inhibit offensive behaviour without excessive recourse to expensive and uncertain juridical judgments.

Here, the balance of advantage seems to us to lie with Justice's formulations as embodied in the Cash Bill, although the first provision, that 'spying, prying, watching or besetting' constitute an infringement of privacy, seems too broad. In terms of international comparisons such a regime, though stronger in its protection of privacy than the current system in Britain, is considerably weaker than the regime available in France, Germany and some Canadian provinces.

A right to privacy can only be a right of natural persons. We are more concerned about ordinary people, whose private doings are made public, than about public figures, whose tribulations seem to be uppermost in the minds of regulators. The claim for protection of ordinary people, such as the teachers[12] whose lunchtime activities were salaciously reported in the *News of the World* (15 October 1995), we find more compelling than that of public figures. And, perhaps more contentiously, we believe it is right that persons who have chosen to put themselves in the public eye, or have accepted roles and occupations which have a public character, should be less likely to succeed in claims of infringed privacy. This was the approach taken by the US Supreme Court in its famous judgment in 1964, *New York Times* v. *Sullivan*, which found that even erroneous statements about a public official were lawful. It may be that the Supreme Court opened the door too wide to the publication of falsehoods about public figures (requiring plaintiffs to show that there was either knowing publication of falsehood or reckless disregard of the truth or falsity of the publication) but its emphasis on the overriding importance of the availability in the public domain of expression about public figures is surely correct (see, *inter alia*, Barendt 1991).

However, the division between public and private personality is not always clear cut. Those such as police, teachers, social workers and priests, who discharge public duties and in whom the public repose trust (as it must if those charged with such duties are to discharge them successfully), will enjoy less protection of the privacy of their personal expression than will those who have no such public roles. The legitimacy of infringement of privacy must always be in proportion to the public interest and not in response to what the public is interested in.

Rights are seldom absolute. Both the right to privacy, if right it be, and the right to freedom of expression and information are limited by considerations of proportionality, public interest and the claims of countervailing rights. Privacy, freedom of the press and freedom of expression are irrevocably intertwined. The exercise of the right to expression may sometimes clash with the right to privacy, but new laws on freedom of expression and privacy can act together to strengthen realization of both rights.[13] However, the possible opposition of rights to privacy and rights to expression points to the importance of those who decide how to strike the balance. Who decides?

Broadly, there are two regulatory alternatives in the UK: the statutory model exemplified by the Broadcasting Complaints Commission, and the self-regulatory model exemplified by the Press Complaints Commission. Whether acting under a statutory or a self-regulatory hat, those who decide are not dissimilar people: older, maler, whiter, better educated and more London based than the population as a whole (see appendix 2). The obvious remedy is to make the decision makers more representative of those for whom they decide. In chapters 7 and 8 we suggest how regulators should be appointed. Here it is worth noting the importance of consultation with the public and of periodical public deliberations and discussions of the principles and practices of regulation.

We propose therefore that the statutory offence of infringement of privacy be established. Complainants who seek only correction of error and apology for infringement of privacy may elect to accept adjudication by the regulator we propose (see chapter 8), Ofcom. The regulator's judgments must be made public and supported by reasoned argument and/or reference to precedent. Where Ofcom upholds a complaint, it shall have the power to require publication of correction and/or apology, and where publications offend persistently, Ofcom shall have powers to fine the offending publication and to publish a report of the judgment and the penalty exacted. Complainants who seek financial redress for infringement of privacy should use civil action for damages. Informal agreement between complainant and defendant should of course be an alternative option open to both Ofcom and judges. Ofcom's adjudications on privacy shall be subject to judicial review (Liberty 1994 pp. 43–4; Curran 1995 pp. 38–9).

Freedom of expression and changing technology

Two mistakes are easily made when thinking about new technology. The first is to think that new technology means we have to revisit all our principles. Instead, we can apply our principles to each technology. The more public and intrusive the medium, the more stringent the restrictions. Conversely, more controversial material is available on media that enable restriction to a consenting adult audience.

Table 5 ranks media in ascending order of 'privateness'. At one extreme is the telephone: a private form of communication, restricted to a volunteer audience, where passwords (or PIN) can be used to exclude unauthorized users from unsuitable services. At the other end is the poster, where none of these conditions apply. In the middle are question marks where reasonable people may disagree. Here, the uncertainties about the standards of reasonable people require regulators to consult with the public – perhaps through citizens' juries. To obtain the same freedom as the telephone, a technology must demonstrate the same characteristics: a volunteer audience, restrictable to adults. Whether or not a technology is a mass medium has little effect in this equation. For example, it would be as offensive to distribute racially inflammatory material to one person as it would to 100,000.

The second easily committed mistake is to think that because regulation is not watertight we should abandon the attempt to regulate at all. New technologies like the Internet do raise new problems, such as jurisdictional disputes over messages mailed in one country and read in another. But most supposedly new problems exist in current technologies, and are simply reproduced on a larger and faster scale over the Internet. But the fact that we cannot detect and punish all crimes does not invalidate the attempt to do so: just because we do not catch all murderers does not mean that we do not prosecute those murderers that we do catch.

Those who say the information superhighway marks the end of censorship fail to distinguish between suppression, restriction and punishment. It is going to be increasingly difficult to suppress what we want to suppress. But we will still be able to restrict material to mostly volunteer, adult audiences, and still be able to prosecute where there is a victim.

Table 5

	Posters	Free-to-air television	Press	Subscriber television	Videos	Video-on-demand	Telephone
One to one	no	no	no	no	no	?	yes
Volunteer audience	no	no	?	no	yes	yes	yes
Restrictable to adults	no	no	no	?	?	yes	yes

Moreover, there is a happy coincidence of public and private aims, since superhighway service providers will want to assure users that children can be protected and adults not offended. The idea, proposed by ICSTIS, of 'safe havens' on the superhighway free of offensive material seems likely to be realized.

Summary

1   In media, content regulation is important. This chapter examines the foundations of UK content regulation, in order to establish new principles that can be clearly applied to new and traditional media. Increasingly, content regulation will be extended from media to new telecom domains, like the Internet.

2   We argue for rights to freedom of information and expression, which presently are not guaranteed by law in the UK. These rights must have clear boundaries, in well specified circumstances. However clear the rules, there will always be some degree of controversy: those who decide over such disputes must have a public mandate.

3   Freedom of expression should be a constitutional right. The European Convention on Human Rights should be adopted into UK law. It still provides discretion for the state, which should be limited in the UK.

4   The principles of the Williams Report should be used to build our right to freedom of expression and as a general basis for content regulation. It would yield the following general rules:

   (a) *Suppression before publication* can only be enforced in cases of clear and present danger of harm to national security or to public order. The defendant is allowed a public interest defence.

   (b) *Restriction to an adult audience* applies to material that is not appropriate for children and should be implemented through codes drawn up by the relevant regulator. It has the advantage of keeping implementation out of the hands of politicians, who may be more susceptible than regulators to short term moral panics.

   (c) *Restriction to volunteer audience* protects people from stumbling into offence in the course of their day-to-day activities, and hence it applies to public media, for example posters or television. The notion of what is public media changes, and should be kept under review by the regulator. However, the converse is that adult, volunteer consumption should not be censored because another adult finds the material consumed offensive.

5   Freedom of information is another important aspect of free expression. We argue for a more open public sector, for a right to know embedded in a statutory right of access to official information.

6   A right to privacy should be recognized by law. Breach of privacy can be accepted only for public figures and if in the public interest. We interpret the public interest to mean information that is relevant to the public role investigated, not information that satisfies public curiosity about the person.

7   These principles can be applied to new technology, by judging to what degree the new medium allows restriction to an adult or volunteer audience. Suppression is arguably more difficult with new media, but offenders can still be prosecuted.

6 Audio-Visual Policies: Too Much or Not Enough?

The audio-visual industry embraces the production of sounds and images as well as the apparatus needed for their consumption. We concentrate here on the 'software' or 'content', i.e. the production of sounds and images, and ignore the hardware or manufacturing of the various machinery that allows us to 'play' our chosen software. Nevertheless, we recognize the important links between the two segments of the market. Any new hardware product successfully launched into the markets will initiate a wave of new software production and affect the operation of existing software markets.

Film production is at the core of public concern about the audio-visual industry. The importance of the film making industry can be stressed from many different points of view. From an *economic* viewpoint, making films is a good way to create jobs and increase exports, as the industry is highly dynamic (world markets for entertainment grew by 400 per cent in 1994). Entertainment – largely Hollywood's film business – is the United States' second largest export sector (*New Media Markets* August 1995). This brings about a second popular argument for promoting cinema production. Hollywood's exports produce one of the most significant trade deficits in the EU. The *trade deficit* argument is advocated strongly for imposing special measures to restrict imports and stimulate domestic production.

From a *cultural* perspective, films help to create a nation's heritage of shared values, images and history. They are a powerful way to portray the present and past life of a country and its people, to celebrate traditions and expose problems and moods. Cinema production is the modern way of creating tales and telling stories and is a fundamental component of the artistic wealth of a nation. Films also offer an important platform for social issues to be brought to the fore. Films are a cohesive stage where different voices can emphasize their differences, common values can be exposed or reinforced, marginal cultures can forge role models with the same resonance as mainstream ones. Films enable minorities to express their values and illustrate their common heritage. Dramatic productions can help to cement a multi-cultural society to everybody's advantage.

Public policy is rightly concerned with the well-being of the national film industry for all these reasons. But those who argue for special treatment for the film production sector do not all advocate the same strategies.

Some stress the need to establish a commercially and internationally successful *audio-visual* industry, not just the film industry. A motion picture is one of the ways in which to explore and portray a character or a situation. Today's film sector is no longer separable from the other types of audio-visual media, and revenues flow from the ability to exploit an original idea in as many formats as possible. Limiting industrial policy to the film industry fails to acknowledge the most important trend – convergence of previously separated sectors – and misses the windows of opportunity that this opens.

Others argue that dramatic productions should not be constrained by commercial requirements and should be able to cater for the niche audiences that may not be economically appealing. Public service broadcasting responds precisely to the need to provide entertainment and drama to enhance national heritage and values, without the restrictions in scope that arise from market pressures. Public broadcasters are mandated to cater for minority tastes and interests, and provide diverse content. Does the cultural argument justify government intervention over and above public broadcasting? We are not persuaded that it does.

Promoting commercial film making requires a different set of policy tools and resources from those needed to foster films as an artistic expression. The artistic value of a film is a matter of judgement which may or may not coincide with market success. Some music considered to have high artistic value has access to national stages and to music shops, but is nevertheless at the margin of commercial production. So too may 'art house films' be at the edge of the commercial circuits, but be an important element of national culture. Supporting artistic, niche film production is an important task, but it does *not* equip a national film industry to compete with Hollywood. Cultural policies are not part of an industrial policy. In the rest of this chapter we will analyse the wider audio-visual sector at large, and focus on *industrial* policies for fostering the commercial success of the sector.

EU film policy

In the last decade, US studios have been instrumental in reviving European cinemas. They have entered the exhibition sector, building multiplexes in many countries (UK, Germany, Spain), thus helping to reverse the declining trends in European cinema-going. They have also steadily increased their European revenues and their huge audio-visual trade surplus with the EU, which in 1993 stood at $3.7 billion (*Screen Digest* July 1995 p. 168). Table 6 shows the share of US films in EU markets in 1992: about three-quarters of European box office revenues go to US producers and distributors.

The television industry has seen an explosion of demand and rising software prices since the growth of commercial broadcasting and pay television. To give an idea of the scale and rapidity of this transition: Europe had four commercial television channels in 1982 and fifty-eight in 1992 (Hodgson 1992). By January 1996, there were 217 commercial channels in Europe (*Screen Digest* January 1996 p. 9).

A large share of this growth has been absorbed by US programmes. International prices for US television drama – which have already recouped a large proportion of their production costs within the US domestic market – are easily ten to twenty times cheaper than equivalent

Table 6   Share of 1992 EU film markets by country of origin

| Market | Percentage box office for films of origin: | | | |
	US	Domestic	Other EU	Other
Belgium	73	4	19	4
Denmark	78	15	3	4
France	58	35	4	3
Germany	83	10	6	1
Greece	93	2	3	2
Ireland	88	8	4	0
Italy	69	19	11	1
The Netherlands	79	13	3	5
Portugal	85	1	9	5
Spain	77	9	13	1
UK	86	12	1	1
European total	74	17	7	2

Source: London Economics and BIPE Conseil, *White Book of the European Exhibition Industry*, 1994, vol. 2, Milan, Media Salles

original EU drama. Furthermore, very few EU productions can effectively be considered 'equivalent' or substitutes for US drama, given that the main (perceived) weakness of European producers is their inability to create material which appeals to international audiences. As a result, there is a shortage of programmes attractive to audiences across Europe, and programme circulation within Europe is also very low.

The European Union expressed concern about the inability of its own film makers to compete with the US, and began devising policies to revive the sector and improve the imbalance. The creation of a European marketplace for broadcasting was the main objective of the Directive on Television Without Frontiers (Council of the European Communities 1989) which was designed to harmonize broadcasting regulation across Europe and promote the European television market. Among the measures introduced by the Directive is the requirement to transmit a majority of European productions, discussed below.

European audio-visual policy also introduced a series of initiatives, collectively known as the MEDIA programme, aimed at strengthening national EU audio-visual markets by encouraging greater co-operation and better distribution of their products. The first phase of MEDIA spanned from 1991 to 1995, with a funding of approximately £140 million. The European Film Distribution Office (EFDO) was created in 1988, with the aim of providing help to medium and small budget European films for wider European release. EFDO's assistance takes the form of soft loans, to be repaid with box office receipts, when distributors have recovered their shares of pre-costs and overheads.

The White Paper on *Growth, Competitiveness and Employment* (Commission of the European Communities 1993) earmarked the audio-visual industry as the growth sector of the next decade, where most new employment will be created: it is expected to provide over 2 million jobs in the next few years. The White Paper gave new impetus to EU initiatives aimed at the communications sector at large. A second phase of the EU audio-visual initiatives, the MEDIA 2 programme, will start in 1996. The most important change is the Commission's proposal to create a $260 million guarantee fund as a financial instrument to stimulate private sector backing for cinema and television production. The structure of the programme has been grouped along three principal action lines – distribution, development and training – to be administered by three intermediary organizations independent from the Commission and selected by competitive tender.

The EU has been active in promoting the creation of a unified European marketplace, but much still needs to be done. The European film industry is far from being integrated: it is fragmented into national markets. This weakness is compounded by a low rate of cross-border programme distribution and circulation. As table 6 shows, non-domestic European films were able to get only 7 per cent of the EU box office revenues in 1992. Small producers are unable to compete in European and world markets, and their own survival is often in danger. The industry is trapped in a chronic deficit spiral and is unable to attract European capital, although the same capital is invested in Hollywood ventures.

The 1994 Green Paper on *Audio-visual Policy in the European Union* (Commission of the European Communities 1994a) is principally concerned with the film and television industries. Their role is believed to remain central because technological changes, such as digital compression technology, are set to transform the sector, accentuating the strategic role of the programme industry. The Commission identifies four fundamental requirements for the EU audio-visual industry:

1 It must be competitive in an open world market.
2 It must be forward looking and be involved in the development of the information society.
3 It must illustrate the creative genius and the personality of the people of Europe.
4 It must be capable of transforming its growth into new jobs in Europe.

Safeguarding the diversity of national and regional cultures is now clearly linked to the development of a predominantly European programme industry which must ultimately be profitable. European policy is gradually shifting away from a 'black box' approach which assumed that, if subsidies were poured into the production sector, more and better productions must follow; it is now placing greater emphasis on supporting and strengthening existing winners, within a general competitive framework. The Green Paper employs two distinct approaches to financial incentives:

• Support a maximum number of programmes, to give as many talented people as possible a chance, with the aim of increasing diversity.
• Give priority to activities which strengthen the base of companies with potential for a number of programmes, thus spreading the risk (1994a p. 43).

The first set of measures appears to suffer from the same flaws as the black box approach, that is to say the lack of incentives to enhance efficiency. It does not help the industry to adopt a more commercial appeal, nor does it guarantee diversity. Wider access to production is indeed a crucial starting point for the promotion of diversity, but a general commitment to supporting many programmes could end up doing nothing to improve access. It may well end up supporting all sorts of projects, but only benefiting those who already are involved in the industry.

The second approach reflects the new thinking of the EU, which tries to build upon existing strengths, rather than creating new ones from scratch. It looks at the overall activities of companies and their success in the European market, or provides incentives to groups of companies.

This dual approach underlies a basic confusion which seems to affect most debates on the cinema industry: policy makers argue that protection of the EU's cultural heritage is the main reason to give special treatment to the film *industry*. If the development of an industrial sector is the ultimate goal, policy must rightly be concerned with issues of efficiency, and with ensuring that the relief it grants to the industry is temporary and targeted at a specific problem. In his intervention at an EU conference, a speaker from Polygram, one of the major European producers, expressed this concern by stating: 'My point is that there is a distinction to be made between international and local product, between Hollywood and local movies, between a business which should suffer no restrictions and be given no special treatment and a business which will only survive with special treatment' (European Audio-Visual Conference, Brussels, 30 June to 2 July 1994). Emphasis is now correctly being placed on the creation of an industry capable of supporting itself. Policies to support cultural productions are different: they do not need to be overtly worried about efficiency issues, and should remain distinct.

Is anything wrong with quotas?

The most controversial aspect of the Television Without Frontiers Directive is the imposition of a programming quota restricting broadcasters to scheduling *wherever practicable* – no more than 49 per cent of productions of non-EU origin (Article 6). Alternatively, broadcasters can choose to re-invest a set percentage of their budgets in local productions. Further, at least 10 per cent of the relevant categories of programmes[1] – or, again, of the broadcaster's production budget – must be reserved for independent producers. The imposition of quotas sparked a lively debate about the ability of quotas to deliver the final objective: more quality European productions at prices closer to world prices. At the time of writing, the final text of a revised Television Without Frontiers Directive is close to being finalized. In November 1995 the Council of Ministers voted to confirm the current provision of Article 6 – 51 per cent of European quota *where practicable* – and extend it to stock programmes, that is to say programmes with a shelf life.[2] However the European

Parliament has voted for stronger provisions and the definitive content of the Directive has yet to be determined.

Trade restrictions which limit the amount of imports from any country, like quotas, or from selected countries, like voluntary export restraints (VERs), offer home producers of the traded good a shelter from outside competition. In theory the price of imports rises, home production increases, and the local industry has a chance to exploit scale economies, learn and become better fit to face outside competition. In general terms, these restrictions are not in the public interest, as they transfer money from consumers (they pay a higher price for the traded goods) to uncompetitive producers, who can then use it to increase efficiency or to increase profits. The link between greater local production and improved efficiency, however, only materializes *if* the price gap between imports and local products is due to learning or scale inefficiencies, which disappear with greater output and as firms climb their learning curves. If, on the contrary, the higher costs are due to other inefficiencies, quotas are likely to be counter-productive, by creating a protected environment where no incentives are given to greater efficiency. This is in fact the single greatest flaw with quotas: producers have few incentives to adjust their creations to audiences' tastes and to contain their costs.

However, theoretical discussions about trade restrictions and the conclusions that quotas are against the public interest are based on an ideal background which is very different from the real world of audio-visual trade. Given the advantages enjoyed by US producers *vis-à-vis* the fragmented EU industry, the argument for temporary measures to cushion the path towards a real EU-wide audio-visual market is worth exploring. To do so we must consider:

- What are the most suitable protection measures?
- What will be the costs?
- What will be the consequences?
- How are the measures likely to be circumvented by those who stand to lose from them?

As far as trade restrictions are concerned, VERs are better than quotas, because the EU is worried about US imports, and VERs target US imports only, whereas quotas restrict entry of all non-EU audio-visual products. VERs are also superior to tariffs, because they transfer revenue to the targeted production sector rather than to the EU coffers as import tariffs would. But they share the same flaw of all protection measures: there is no guarantee that the protected producers will use their additional earnings and time to improve their efficiency.

Examining VERs in the US and Japanese car trade in the mid-1980s showed that the value of trade increased with volume restrictions, and that the national industry was able to compete with Japan after a few years of restrictions. However, the cost to consumers was significant, as US car manufacturers raised the average price of the cars sold in the US rather than improving their productivity during the VER period.

The peculiar features of the film markets make a successful application of VERs even more debatable:

1 US producers are vertically and horizontally integrated and EU producers are not. With a VER, US producers would definitely shift their mix of imports towards the blockbusters, which are the film releases the EU is least equipped to compete with. These would become even more expensive for the (local) exhibitor, without harming the integrated US multiplex exhibitors. Distributors are also likely to raise their margins, and in most cases this means US distributors.

2 The EU market is geographically fragmented. Would EU producers raise their prices in response to a VER? Or would some prefer to try and gain market share by keeping prices low without improving their competitiveness in quality terms? Are there EU countries that stand to gain more than others?

3 Arguably, domestic and foreign cars are more easily substitutable than US and EU films. The lower the degree of substitution, the more likely it is that the overall increase in film prices benefits overseas rather than domestic producers.

4 There is more than one retail market. Films are sold through a number of windows. What will be the impact of higher film prices on the television, video and video-on-demand markets? How will EU broadcasters adjust to a scarcity of cheap US dramas?

Some of these issues are addressed in a document written for Sony by London Economics (LE) (1994c). LE argues that quotas are likely to affect only new entrants to the European broadcasting market, because established television channels, mainly public and terrestrial commercial stations, already schedule at or above quota levels. The idea behind quotas in television dramas and films is that, since (a) audiences prefer local productions, and (b) television channels are the greatest purchasers of films, imposing the quota will ensure that the growth of demand for TV drama and films is not absorbed by US productions. However, new entrants faced with an obligation to buy more EU productions, which are substantially more expensive than equivalent US material, have two options: they can switch to cheap Euro-junk, hence losing audience, or they can raise their costs by buying quality EU productions. Either strategy is costly and may retard development of new EU channels and/or limit audience. Quotas would increase local content only in a few specific cases. LE's argument rests on the assumption that *in time* new television stations will turn to local production – because that is what audiences prefer – and concludes that quotas may considerably delay the time when a new broadcaster can afford to undertake local investments in production. However, the assumption is true of generalist TV, but ignores the new dedicated channels of the 1990s, which may want to leave local production entirely to public or established broadcasters. The very existence of these channels would be threatened by imposition of the programming quota, since their budgets are based on a schedule of cheap American imports. The point here is not to forbid the creation of all-American channels, but to ensure that the advantages enjoyed by American producers (such as the ability to recoup costs in their home market) work in favour of the emerging EU market – so that Europeans too can buy programmes cheaply – rather than against it.

Investment quotas, rather than transmission quotas, would serve the purposes of increasing EU production more efficiently. Rather than limiting the ability of the dominant producers to sell into the European markets, this ability should be made conditional on positive contribution to local productions. Unlike a subsidy, which channels money into production without any consideration for the commercial characteristics of the results, investment requirements force broadcasters to monitor the returns on their investment and thus foster efficiency. Policies that reduce the life chances of TV ventures which are mainly US in content or ownership do not benefit the EU audio-visual industry and the public at large, but the requirement that US investors who make huge revenues in Europe reinvest some of those revenues in the EU undoubtedly advantages Europe.

UK policies

Recent British box office hits, like *Four Weddings and a Funeral*, *The Crying Game*, *Orlando*, *Shallow Grave*, and so on, are reviving interest in the UK film industry. Rather than signalling the restoration of a British film industry, these isolated episodes remind us of the pool of talent that remains underused in Britain.

The UK film industry has been characterized by a declining trend for the past ten years. The number of UK productions was at its lowest level for over a decade in 1993, when only 10 out of 242 pictures on release in UK cinemas were British (National Heritage Committee 1995 vol. I appendix F table 2). Even when the number of films per year increases, the hidden picture is that UK production has moved towards smaller and smaller budgets, and dwindled to the point that most producers are today not able to survive longer than one picture. The average budget of US majors' productions in the UK in 1993 was £11.54 million (British Film Institute 1995 p. 25); UK films cost on average £1.32 million.[3] The proportion of British productions that failed to get a cinema release was 31.9 per cent in 1992, compared to 2.5 per cent in 1983 (p. 30). Investment in British cinema is down, the share of British films in UK cinemas keeps declining, and the existing pool of skills risks depleting for lack of training. Lack of investment in the British film industry means that the sector's survival is threatened and that a vicious circle of low investments, low returns and lower investments will discourage people from entering the business. An industry like cinema, in which 'talent' is the main asset, is quickly deprived of its key people, who go to Hollywood as soon as funds run dry. Replacing skilled workers becomes problematic when the combined pressures of low confidence, high risk and little money discourage young people from pursuing careers in cinema and fail to attract training funds.

The main cause of the present situation is a change in national policies, which not only eliminated incentives to investors but also abandoned a levy system[4] which channelled funds to producers. The adverse impact of these policies was compounded by the upward trend of the exchange rate in the late 1980s: sterling became increasingly expensive relative to the dollar and US film makers kept away from Britain. In 1991, the government set up a one stop shop for international investors, the British Film Commission (BFC), to attract foreign film makers to Britain. The BFC provides a complete database of filming facilities in Britain, and all the information an international company would need when researching the UK as a location for filming. It has generated inward investment in excess of £80 million in its first three years of existence (1991–4) (British Film Commission 1994), but one can only speculate as to what proportion of such investment would not have been directed to the UK without the BFC.

The British Film Institute (BFI) has overall responsibility for the development of the culture of the moving image in all its forms. In addition to its museum, archive and library services, the BFI has several divisions that actively participate in UK production, distribution and exhibition. Its production division has a basic budget of £1 million and gives priority to films whose budget does not exceed £450,000. The BFI is mainly concerned with developing new talents. However, the film strategy of the BFI has failed to engage the support of the production sector and its role is increasingly being marginalized.

The only direct government support to commercial cinema today is a grant of £2 million a year to British Screen Finance, a private company created in 1986 and owned by Rank, MGM/UA, Channel 4 and Granada. British Screen has played a far more important role than BFI in galvanizing British film production. British Screen administers the European Co-Production Fund (ECPF) on behalf of the government (since 1991, with an additional £2 million a year) and provides commercial loans for script development, films and shorts. Government's contribution to British Screen has been unchanged since 1986, which means that the real value of UK public assistance to the industry has dropped by around 46 per cent. British Screen contributes no more than 30 per cent of a film's budget, and contributes mainly to pictures that would not otherwise reach the production stage. Its contribution is very important for individual films, but does not address the frailties of the industry as a whole.

National Lottery funding will be available for film production, as part of the funding earmarked for the arts. Government has reserved Lottery funding for projects involving capital expenditure – especially the construction and improvement of buildings – and for the purchase of equipment. The definition of capital expenditure has been stretched to include film as a capital asset.

The Arts Council identified three main objectives for funding film:

- to increase the overall level of 'British' productions (under the terms of the Films Act 1985)
- to increase the range of investment opportunities and the variety of genres in which investment may be encouraged
- to spread the burden of risk for investors by the provision of secure financial support.

These broad objectives offer considerable scope in terms of the nature of the Arts Council's support for production. At present the pilot project is focusing upon investing primarily in the production of features, rather than other elements, such as distribution or exhibition venues, which could equally boost the film industry.

Although the Arts Council is not allowed to earmark specific amounts for particular uses, it is likely that about 15 per cent of the Lottery arts budget will be allocated to film production. Estimates mark the Lottery funding to the film sector at around £70 million over a five year period. Following early awards by the Arts Council the figure seems closer to £75 million, with £15 million being invested in the first year alone.

Such levels of funding compare favourably with the £4 million per annum guaranteed to British Screen by the government. However, the allocation of Lottery funding to individual projects, still subject to debate, seems to range between £5,000 for low budget and experimental projects to a ceiling of £1 million for low to medium budget, commercial features. Such a ceiling is too low and higher funding to assist financing of medium budget British features (costing between £10 million and £15 million) should be authorized.

Lottery funding will not equip the British film industry to compete with US blockbusters, and is not meant to. Lottery funding is meant to benefit British projects and the British public directly. Commercial features are too high risk investments to be earmarked for Lottery funding. However, this does not explain why the Department of National Heritage directions for allocating Lottery funding have been stretched to include the creative part of film production, without addressing more straightforward structural projects in the exhibition or distribution sector, which could benefit both the industry and the public. Access to non-US products on British screens outside London, for example, is minimal, and more can be done to boost attendances at regional film theatres. Support for distribution and exhibition is likely to be more useful than support for production.

The earmarking of Lottery funding marks a welcome change in the government attitude towards the film industry. Government has repeatedly insisted that Lottery money will not replace existing forms of arts funding. Nor must it be allowed to become a substitute for a coherent and comprehensive government film policy. Regrettably, in the months following the first Lottery awards, the government failed to introduce fiscal measures aimed at encouraging investment in feature films and it announced withdrawal of the UK from Eurimages, the EU co-production agency.

The structural problems of the UK film industry require urgent attention. UK producers have trouble raising finance, and UK films are therefore made with small budgets and small ambitions. Financing UK films is riskier than investing in Hollywood productions because:

- When budgets are small, returns are small and they materialize several years later; distributors and exhibitors recoup their investments before financiers, and in any case not before three to five years.
- Reducing risks by increasing volume of production requires about eighty films per year, and recently there have been fewer than fifty productions each year.
- There are no incentives for domestic or international investors to counter the disincentive of higher risks (strong argument for creating tax incentives).
- The production sector is isolated from other aspects of film making –

namely distributing and exhibiting – and the possibility of spreading risk this way is hence limited.

The links between distribution and production are particularly important for the purposes of raising finance. To generate the cash needed for shooting, a film is generally pre-sold to a given number of territories at advantageous prices. The pre-sale contracts are used to back up cash flow requests to the banks, and the difference between pre-sales and actual production costs is usually met with subsequent distribution revenues. In the words of the experts, 'ideally, a film is sold first, then actually made' (Lewis and Marris 1991 p. 5). Inefficient links between distributors and production agents may imply that films are not sold – *and* not made.

Distribution is a second significant weakness of the UK production sector. Practically all sizeable distributors in the UK are subsidiaries of US majors. The top five distributors[5] – all majors – reaped 84 per cent of total box office receipts in the UK in 1992 (London Economics and BIPE Conseil 1994 vol. 2 p. 116 table 40) and 82 per cent in 1993 (British Film Institute 1995 p. 42 table 19). They have no specific interest in British productions and tend to favour US films, simply because US producers can offer better terms (production costs are likely to be recouped with home distribution) and better marketing. The *White Book* states that:

> There is evidence in the UK, Germany, and Italy that US distributors make more from the EU films that they distribute than from US products. This indicates that US distributors may well be biased towards US products as we would predict, because they need a higher return on European products to achieve the same margin as on their own products. US distributors will distribute films from other countries, but they expect them to exceed average performance of US films. Since US distributors dominate the market, this means that access for independent productions may be limited. (London Economics and BIPE Conseil 1994 vol. 2 p. 118)

Even the recent improvement in the number of screens has not been translated into more UK films getting released.

Broadcasters are also key investors in production. They realize a return on their investment, however, if they fill a slot in their schedule with a product that audiences and advertisers will like. They have a limited interest in promoting productions for cinema release, which is where the perceived value of a film increases, and other windows. Television channels usually retain all the rights to the programmes they acquire, and they are able to dictate producers' operating margins. Independents who retain some of the rights to their productions may be able to maximize revenues from alternative distribution and exhibition channels. They can thus become more stable financially and truly independent of broadcasters' demands.

New technologies – video, satellite and cable televisions – were expected to provide additional demand for local productions, but have so far contributed only minimally. BSkyB, for example, has struck a deal with British Screen Finance by which it invests £2 million each year in return for television rights on all British Screen productions. For the first few years, however, like all other new entrants into broadcasting, BSkyB has survived on a diet of imports. Videos have now surpassed cinema attendances in revenue terms, but their contribution to production is very limited. Many in the industry call for some form of levy which would pump money from videos to studios.

In summary, the main weaknesses of the UK film industry are:

• high risk and no fiscal incentives for investors
• limited public finance
• uninviting terms and conditions for foreign investors
• no integrated producer/distributors
• no significant local player in distribution
• fragmented independent sector with over-capacity
• inadequate regulation of rights for commissioned drama
• low levels of training
• low levels of investment in production from new broadcasters and alternative windows of release.

Many different solutions have been advocated to address these issues, and some of them are corrected by EU initiatives, such as MEDIA 2, which will fund training across Europe. However, the present policies do nothing to address the structural weaknesses of the UK film industry; they only provide a lifeline to keep a collection of projects going.

The industry needs to break out of the 'small budget' mentality and look at the international market in order to start achieving sufficient revenues to sustain itself. The film industry cannot be seen in isolation from the wider spectrum of audio-visual content. Creativity is shared among the different segments of the market. Revenues can be increased if risk and all common overheads are shared among a range of audio-visual products. Producers who are not fighting for their livelihoods would have all the incentives to integrate vertically or horizontally, but an industry on a lifeline cannot be expected to have the time and resources to devote to strategic planning.

Policy must address the bottlenecks at the supply side of the core industry, and provide fiscal measures to encourage investment in films. The success of measures undertaken by the Irish Republic (tax incentives for location of films in Ireland) shows that international investors can bring substantial funds to the national industry if the right incentives are in place. But all the initiatives set up to improve film distribution, the marketing of films in different formats and windows of release, and the regulation of rights, should not be restricted to cinema fiction. Nor should they impose a hierarchy of delivery systems where, for example, operators choose the type of release that would maximize tax allowances – say, cinema – instead of the release tactic that would maximize the film's total earnings – say, straight to video.

We recommend a revision of the fiscal regime of film production, with the aim of providing incentives to invest in film making and removing any other differences in the treatment of filming *vis-à-vis* other audio-visual productions. Present support for the film industry is only adequate for small scale niche productions, and more money should become available with a combination of fiscal incentives, European funds and minimum reinvestment requirements as set in European legislation.

When delivering commissioned work, producers should be able to retain rights to exploit their work in alternative formats, particularly while the market has a plethora of small production houses gravitating around a handful of commissioning broadcasters. Public money – Lottery money – should be earmarked for public exhibition sites like the regional film theatres, which bring greater choice and more European cinema, outside London. The availability of state-of-the-art cinema theatres, multi-media complexes and film libraries can do a great deal to promote greater interest towards an alternative to Hollywood cinema. But providing the right infrastructure cannot help British and European cinema unless producers deliver the entertainment that audiences have learned to expect from motion pictures.

New Media

New media encompass a varied range of products, with these characteristics in common:

- They are screen based.
- They can offer images, text and sounds.
- They allow some form of interaction.

Most of the world's leading audio-visual producers are trying to establish their position in the market, by launching their own system to play multi-media products. A few competing platforms are emerging, some of them based on the television screen, such as the very successful

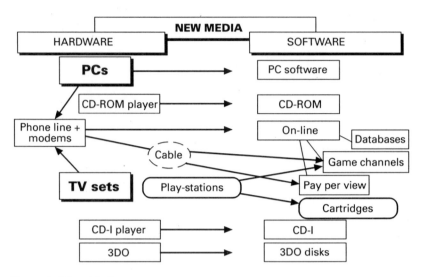

Figure 4   New Media

play-stations, and some on the personal computer (PC). Figure 4 shows the range of choice in hardware and software currently available.

The struggle for a winning system must take into account both demand and supply conditions, that is both what is cheaper to produce and what consumers are more likely to want. Television sets are a better platform for a wider and faster pick-up of new media; they have the advantage of a higher penetration rate (98 per cent of UK households). But the installed base of computers is encouraging: one in seven UK households owned a PC in September 1994 (*New Media Age* 17 August 1995). In the US, where nearly one in two homes (46 per cent) own a computer (*Newsbytes* 19 August 1995), 8 million PCs are already connected to on-line interactive services. PCs' important advantage over television is that they are used for interactive purposes, whereas television is regarded as a source of relaxation, and viewers seem reluctant to use it interactively. PCs are presently bought by richer households, keen to spend on education and entertainment. Also, according to a *Financial Times* management report (Adamson and Toole 1995), multi-media are likely to develop via the PC, rather than the television set, because it is easier to improve image quality in a PC than it is to find a cheap and reliable way to make a television set interactive. In time, however, the distribution method is likely to become irrelevant.

Consumer trials of on-line systems to date indicate that games are most used, followed by home shopping, video-on-demand and retail banking. Although American entertainment conglomerates have a large absolute advantage in their content libraries of films, TV drama and cartoons, UK content makers have a very strong position in many other genres. UK banks like Barclays and NatWest are participating in the largest interactive trials and preparing to offer on-line banking facilities. The shopping channel QVC (UK) made revenues above projections for the first time in the first quarter of 1995. The UK media group Flextech joined forces with Quantum International to launch an instantly profitable Sell-a-Vision infomercial channel from Astra (*Screen Digest* July 1995 p. 160). UK Living and the new channel TVX also began infomercials in 1995. In 1995 UK firms provided 15 per cent of all written CD content and 35 per cent of the international property rights for all games (*Computing* 19 October 1995 p. 22). The UK is only at the beginning of exploitation of its potential for software production.

Video games

The video games industry is the biggest success story of the last decade in the consumer electronics market. The market leader, Nintendo, made more profits than Microsoft in 1993, and sales of video games have reached unprecedented levels, making up half of sales in the UK Virgin megastores. The retail value of the UK video game industry in 1994 was £603 million, or more than 20 per cent of the European market (Durlacher 1995 pp. 48–9). Table 7 shows world sales of both video and computer games.

The industry is dominated by two players, Nintendo and Sega, who have 95 per cent of the world market: 97 per cent of the console market

Table 7 World-wide software sales

	Total software sales (£ billion)	Games software sales (£ billion)	Games as % total
1990	33.75	6.75	20
1991	38.03	9.89	26
1992	41.51	13.28	32
1993	48.75	13.16	27
1994	54.21	13.55	25
1995 (forecast)	58.49	14.19	24

Source: Durlacher 1995

and 70 per cent of total world electronic games turnover (p. 3) The duopoly has meant that consumers were never faced with difficult choices of competing video standards and enjoyed consistent global product support and assistance; indeed according to the *FT* multi-media report the duopoly 'has been instrumental in generating the massive growth of video consoles' (Adamson and Toole 1995 p. 18). Software is both developed in-house and commissioned under licence. Sega and Nintendo use licence agreements to control software quality and maintain strong branding. They control distribution of all software that can be played on their machines, retain distribution rights to all associated software and demand licensing fees of about 30 per cent. Technology cycles are very short and steep, with huge revenues at the peaks. Fast paced technological improvements allow consecutive product cycles, together with strong branding and high cash position, and make this market virtually impenetrable for outsiders.

Forty per cent of Sega's games – by revenue – are authored in the UK (Office of Science and Technology 1995a p. 27). Britain has a strong base of animation and software developers, who can currently access the key mass market only by producing for the Sega or Nintendo play-stations. Opportunities for British software developers could increase if there were an easier access to hardware, i.e. once Sega and Nintendo move from cartridge based to CD based video consoles. Such an opening is likely, given the significantly lower costs of producing CDs: CD based games will cost publishers around 45p per game as opposed to £14 per cartridge (Durlacher 1995 p. 8). Furthermore, the cost of recoding a successful game in order to reproduce it on a different platform is much lower with CDs. The move towards an open CD system will shift bargaining power away from the distributors and towards the game developers, who will be able to offer their product to other buyers. However, given the awesome cash position enjoyed by the duopolists, the absorption of small companies engaged in game development in a vertically integrated structure (from creation/production down to distribution) is the likely outcome. In fact, just like movies, video games are getting increasingly expensive and sophisticated to produce, so that a few successful games have to pay for increasingly expensive flops. And just like the film industry, a strong distribution network is as necessary for success as are creative skills.

Branding and distribution systems are one of the keys to success in this sector. Entering the educational game market could be the best platform for independents to establish a strong brand among the consumer groups that are most interested in the product. A strong public sector drive to invest in children-friendly educational software would undoubtedly stir a market where capacity is too fragmented to compete successfully. Incentives for small companies to invest in brand development and distribution will help. Until more money is invested in promoting and branding UK creative talent, it will be sold, unnoticed, under some other label. And as long as this is the case, the value of UK creators' work will disappear overseas without building the UK industry.

The Internet

The Internet is an informal web, linking computers world-wide. Information is shared using common protocols, or instructions on how to handle the travelling bytes. The Internet is a relatively young creature, born in the late 1960s from military parentage and developed as a politically anarchic structure. Initially its members were predominantly academic (33.4 per cent in January 1993), but commercial operators have now taken over, with 27.1 per cent in 1995, against 23.4 per cent academic (*Screen Digest* April 1995 pp. 81–5). Take-up rate of the Internet exploded in the 1990s with the development of the World Wide Web (WWW), a web of computers within the Internet which simplified access to a huge volume of information. Information on the WWW is accessed through easy-to-use graphical pages which link items through hypertexts, so that any topic is accessible with just a click of the mouse.

Figure 5 shows the growth of the Internet since 1991. This is measured by counting the number of computers registered for e-mail addresses. The real number of users is impossible to define, but a rough estimate

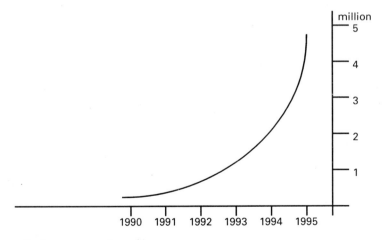

Figure 5   Internet: number of host computers
Source: Network Wizards, Internet Society, *Screen Digest*

assumes that there is an average of five users per (host) computer. However, in most computers with e-mail addresses fewer than half the users have full access to the network (*Screen Digest* October 1994). The rate of growth shown in figure 5 exceeds that of any previous communication or electronic instrument. Growth in Western Europe has been faster than in North America in the last three years, although the USA still constitutes the largest component of the net. Western Europe had over 1 million registered Internet sites in January 1995 (*Screen Digest* April 1995 p. 81), mainly in Germany and the UK.

A number of commercial services provide customers with access to the Internet. The leader in 1995 was America On Line, with 4.5 million subscribers, followed closely by Compuserve (see table 8). These statistics are indicative but do not provide any information on the volume of information accessible at each site, the intensity of usage or the content of information sent and received.

The fortunes of this emerging sector are likely to be shaped by Microsoft, the software giant, which bundled Internet software in its 1995 edition of the PC operating system Windows. Given that an estimated 50 million Windows packages are used world-wide, Microsoft certainly has the ability to dominate and mould the emerging on-line markets. With Microsoft set to become the world's most powerful gatekeeper on the net, other powerful players are teaming up to secure a slice of the promised cake. Cable companies and Hollywood studios are establishing themselves as content providers on the net. In the UK there are currently some forty Internet access providers (*New Media Age* 17 July 1995). Some of them are integrated content providers, like News International's own Delphi, which merged with MCI to provide the next generation of Internet services. Others, like Demon, initially targeted the cost-conscious, cheap and cheerful end of the market, are now fast expanding into Europe, in Demon's case in a venture with Italian operator Europe On Line and with points of presence in Amsterdam and

Table 8　Internet: on-line subscriptions, December 1995

On-line services	Subscribers (million)
America On Line	4.5
Compuserve	4
Prodigy	2
MSN	0.6

Source: companies' information

Paris. Developments of a cable television route to the Internet are slowed down by the lack of agreed standards, which means that each cable operator is developing its own. Multi-media messages and services are technically possible and already available, but their actual fruition still requires a lot of complicated work – and time.

With one of the most advanced and lowest cost telecommunications infrastructures in the EU, the UK is capitalizing on a Europe-wide advantage. The UK is the largest European market for Internet services, at 30 per cent of the $41.6 million market (Ovum research quoted in *Screen Finance*, 28 July 1995). So far, however, commercial applications on the Internet are minimal and at an embryonic stage.

New media – CD based productions, video games killer applications and on-line services – represent a window of opportunity for the UK/EU software industry. They offer additional formats and platforms on which to build economies of scope and exploit potential creative talent. That is why a comprehensive *audio-visual* policy rather than just a film policy at EU and UK level is required.

Summary

1 Industrial policies aimed at enhancing the commercial competitiveness of the audio-visual sector (content or software aspect only) are a tool distinct from and additional to cultural policies.
2 New media, such as video games and on-line services, open new windows of opportunity, i.e. exploitation, for audio-visual productions which need to be addressed in a comprehensive policy framework.
3 UK policy cannot be seen as separate from EU audio-visual policy.
4 General subsidies and trade restrictions fail to address the structural weaknesses of the industry. The MEDIA 2 programme has correctly identified the three main directions – distribution, development and training – that EU policy needs to address. But more emphasis needs to be placed on measures aimed at creating an efficient self-supporting industry and developing the audio-visual sector as a whole.
5 We prefer investment quotas to trade restrictions, whether general quotas or more selective VERs. Investment quotas represent a direct and positive contribution to local productions and promote efficiency.
6 A comprehensive UK audio-visual policy cannot be postponed. Specific measures should aim at providing incentives while removing any differences in the treatment of filming *vis-à-vis* other audio-visual productions.

7 Present support for the film industry is only adequate for small scale niche productions, and more money could be made available with a combination of fiscal incentives, European funds and minimum re-investment requirements as set in European legislation.

8 When providing commissioned work, producers should be able to retain rights to exploit their work in alternative formats, particularly so long as the market has a plethora of small production houses gravit-ating around a handful of commissioning broadcasters.

9 Lottery money should be better earmarked for structural projects such as public exhibition sites like the regional film theatres, which bring greater choice and more European cinema, outside London. The availability of state-of-the-art cinema theatres, multi-media complexes and film libraries can do a great deal to promote greater interest towards an alternative to Hollywood cinema.

7 Public Service Broadcasting: a Better BBC

Broadcasting is changing and the BBC must change with it. The government's recognition of the economic importance of the media is manifest in its White Paper *The Future of the BBC* (Department of National Heritage 1994). The White Paper acknowledges the importance of the BBC and suggests that the government now realizes that the broadcasting sector merits an industrial policy.

However the White Paper's proposals risk changing the BBC from a public service organization (which is at best unaccountable and at worst the cat's-paw of government) to a commercial organization which is similarly unaccountable and vulnerable to illegitimate government influence. They fail to reconcile the BBC's public service role with its important commercial role, nor do they ensure that the preferences of the BBC's users and public are taken into account. They leave the governance of the BBC vulnerable to capture by government.

There are no reasons why a publicly owned, public service broadcaster cannot be commercially successful. Indeed it is desirable that it should be. However, there are difficulties requiring careful institutional organization and engineering, in ensuring that the balance struck between public service and business success reflects the needs and interests of users and the public at large. An organization so widely perceived to be of prime importance to a democratic United Kingdom, as is the BBC, must be accountable to the public it serves *and* independent of other major power centres – whether these be political or commercial.

The BBC: a publicly owned service

The choice between public and private ownership should be pragmatic and determined by which form of ownership delivers the best service.

Telecommunications provides a good example. Before the privatization of British Telecom, telecommunication services in the UK were exclusively provided by publicly owned organizations. The post-privatization improvements in quality and cost of service of British Telecom are discussed in chapters 1 and 2. These improvements are often taken as *prima facie* evidence of the superiority of private ownership. However, this ignores the successes of public ownership of communications, from Hull Telephones' success in providing high quality service at low cost, to the broadcasting successes of Channel 4 and the BBC. Public ownership neither guarantees nor obstructs the provision of good quality (tele)communication services at an affordable cost. The question is thus whether private ownership is the best solution in particular cases. There are powerful arguments for public ownership of public service broadcasting, of which the BBC is the most important – but not the only – UK provider.

Three questions must be addressed in considering the future of the BBC:

- Will broadcasting itself survive the present market trends and technological changes?
- Should public ownership survive?
- Can the licence fee, through which the BBC is currently funded, survive?

Will broadcasting survive?

In one sense the BBC was a technological accident: publicly owned broadcasting was appropriate in the past because there could be no entry to the broadcasting market. Radio spectrum scarcity made a market in radio and, later, television impossible. A non-market form of organization was therefore appropriate. Now entry into broadcasting is possible and *broad*casting itself is arguably drawing to an end. The communications revolution means that hundreds of *narrow*cast channels specializing in particular programme types, such as news or sport, will constitute a plethora of niche markets. The viewers of the future will be charged only for the programmes they watch. They could even order and be sent programmes on demand: instead of waiting for a favourite programme to start, they will choose any programme they want and have it sent to them individually and immediately. Given the ability to select and pay for individual programmes, why would anyone continue to use generalist broadcast channels, paying for a majority of unwatched, unchosen programmes? And therefore why would anyone need the BBC?

Faced with this forceful argument, broadcasters and pressure groups sometimes feel obliged to doubt the communications revolution. They predict slow penetration for these new systems and rightly point to the failure of some current trials. They argue that broadcasting will survive because the communications revolution will not come. But the case for a publicly financed BBC does not rest on technological scepticism. And the communications revolution is already here. It started a decade ago with the transmission in 1982 of Sky Channel from the European Space Agency's Orbital Test Satellite. Since Sky's initial service, the number of commercial television channels in Europe has risen to 217 (*Screen Digest* January 1996 p. 9). The schedules of British terrestrial channels have been extended, increasing the amount of terrestrial television by a third. Both BT and the cable companies plan to introduce video-on-demand; the Société Européenne des Satellites (SES), the owner of the Astra satellites from which BSkyB's services are transmitted, is to provide 500 (digital) satellite channels.

But this technology bonanza will not determine the future of broadcasting: consumers will. Many anticipate that consumers will continue to enjoy broadcasting *as well as* new interactive delivery mechanisms. In the United States, where the modern cornucopia of services is firmly established, the three main terrestrial broadcast television networks continue to account for most viewing. As Nicholas Garnham argues:

> Experience in the US market, a richer market with a mature multi-channel cable market, shows that typically consumers only use with any regularity six channels, that ownership of both the cable systems and the programmes delivered over them is highly concentrated and increasingly vertically integrated, that the classic, over-air, generalist networks retain a significant audience share (currently 60 per cent) and that 40 per cent of the population is still excluded from the cable system whatever its benefits. (1994 p. 15)

The evidence from the first ten years of the communications revolution is that new media supplement, rather than replace, the old. Viewers continue to want schedulers to choose most of their viewing for them, and use new media to supplement core viewing – satellite and video in the 1980s, perhaps video-on-demand in the late 1990s.

The future is unknowable and so predictions, whether of the end of broadcasting or of its continued importance, deserve to be treated sceptically. Throughout the history of electronic media, successive technological revolutions were supposed to replace the existing medium. But new media seldom replaced established media completely: radio and cinema survived television; the cinema has revived with competition from video. In each case, the incumbent medium changed, adapting to new circumstances by focusing on its core capacities: news and music on the radio, a night out at the cinema.

If the past is a guide to the future, broadcasters will need to meet the communications revolution by concentrating on enduring core capacities, at least two of which are likely to grow in importance: the flow model of distribution and programme production.

First, some television will continue to be broadcast. Even when programmes can be distributed discretely, consumers will value current broadcasting's *flow model* of distribution, where consumers pay for a schedule of programmes. As television programmes are 'experience' goods, viewers cannot evaluate their characteristics before paying. The need for 'set menus' of programmes will remain alongside the new 'à la carte' possibilities. As the choice of programmes swells, consumers are likely to value increasingly a service provider whom they trust to select and package a flow of products.

Second, the value of *production* capacity will grow, as demand for programmes rises faster than commercial broadcasters' ability to make them. Consultants Booz Allen have predicted that programme hours demanded will swell from 250,000 per year to 650,000 by the end of the century (Rowlands 1994). Revenues will lag behind, even after new subscription and pay per view receipts are included, and will be divided into smaller slices, as the number of channels rises. Publicly funded producers will be uniquely immune to these financial pressures, and will therefore become increasingly precious sources of programmes.

In the next decade, broadcasting will be challenged: it could disappear, but it could thrive by exploiting its core capacities in new circumstances. Government policy should help the BBC prepare for change. Privatization is not the solution: it would undermine the BBC's core capacities. The BBC is trusted to select a flow of programmes partly because it is non-commercial, and thus able to serve users' interests without pre-emption by shareholders. Privatizing the BBC could undermine this 'brand' advantage, and prevent its exploitation in foreign and future markets. Also, a publicly funded production base would not be exposed to the same revenue pressures as commercial broadcasters. The future of broadcasting does not justify abolishing or privatizing the BBC.

Should public ownership survive?

Some argue that technological change has facilitated effective competition in broadcasting and the public interest can hence be served through the operation of the market. Accordingly a publicly-owned BBC would be an expensive luxury.

As new suppliers emerge, television and radio markets will become competitive, much like the newspaper or book industries: 'By 2000, it is difficult to believe that Britain will need tax-financed "public-service" television any more than it needs tax-financed "public-service" newspapers or book publishers today' (*Economist* 1994a).

This oversimplifies the case. Bandwidth plenty has made competition possible, but is unlikely to produce a perfectly competitive market. It will simply remove one of the barriers to entry in the broadcasting markets. Others – high capital costs, economies of scale and scope, product differentiation – will remain. Competition will improve efficiency and increase choice, but it will not guarantee quality, diversity or impartiality. Intervention will still be needed to remedy information asymmetry and the imperfect nexus between buyers and sellers of broadcasting.

One method for countering market failure is to enhance competition, through privatization and liberalization, and initially enforce it by regulation. Trunk telephony and gas supply are examples of industries that were previously assumed to be 'natural monopolies', but where competition is now being nurtured. Although controversy persists about the aims and sanctions of the utility regulators (Corry, Souter and Waterson 1994), there is wide agreement that they are the most effective way of countering monopoly in those markets.

Should this method be extended to broadcasting? Should we privatize the BBC, and consider it part of a liberalized, but strictly regulated market? The experience of utility regulation to date highlights some important difficulties:

- Competition does not guarantee the erosion of market leaders' dominance: British Telecom, for example, still controls an overwhelming majority of voice telephone calls.
- Partly as a result, regulators intervene more often and in more detail in the day-to-day management of the firms they regulate than governments ever did.
- Finally, tensions between universal service obligations and cherry-picking competitors remain and are not readily resolved through regulation.

These drawbacks of utility regulation lead us to question whether this model should be extended to broadcasting. Privatizing the BBC would mean placing control of a key instrument of democracy in the hands of its owners: Australia presents a striking instance of this danger.[1] And big private players in media are more powerful than other industries: at least none of the privatized utilities can reinforce its power with its own newspapers or television stations. Secondly, the experience of the 1980s undermines our faith in competition policy in broadcasting: it is too easily sacrificed for favourable political coverage. Unresolved regulatory problems which may be tolerable in other utilities would be acute in the case of broadcasting. The beneficiary effects of further competition are a policy aim, but a different regulatory tool is needed. Public service, not-for-profit broadcasting, and the BBC in particular, is that tool.

The 'utility model' of regulation is an example of *negative* regulation: preventing and forbidding certain actions. Media policy is strewn with examples of ineffective negative regulation, including quotas and programme standards (Graham and Davies 1992). As Andrew Graham and Gavyn Davies argue: '[Negative] regulation may be used to require broadcasters to meet a quota for certain kinds of programmes (e.g. news at peak time), it may stipulate standards (e.g. about language), it may even enjoin quality as an aim, but it cannot on its own, ensure that quality actually occurs either within individual programmes or across an evening's schedules' (1992 p. 192).

The BBC acts in effect as an instrument of *positive* regulation: encouraging certain types of behaviour. Through its presence, the BBC provides an important benchmark of standards below which competitors' services sink at their own peril. In a multitude of ways, from technical standards to the development of innovative programme forms, the very existence of the BBC compels other broadcasters to maintain and improve the quality and character of their own services. This has a positive impact on all television and radio consumed in the United Kingdom. As long as the BBC remains a powerful competitor in UK broadcasting markets its services will offer benefits to all viewers and listeners – even those who do not consume them.

A regulator cannot have the same effect as a channel controller, expert at the job of commissioning quality, diverse and impartial programmes. Intervention and regulation are costly: an efficient BBC will be a cost effective solution that minimizes the risk of failure.

Can the licence fee survive?

Could public funding of broadcasting cease to be viable? The would-be privatizers point out that in satellite households the BBC's audience share is already below 30 per cent: its share of the total UK market will continue to fall as satellite and cable spread and the quality of their product improves. They forecast that the BBC's share will hit a 'legitimacy barrier', below which public funding will no longer be defensible. Those who can opt out of the BBC, i.e. subscribers, will have to be allowed to do so. Better to take the positive decision to privatize the BBC now, than be forced to do so in defeat later, when the corporation's brand name and market share have suffered. The Canadian experience lends support to such arguments. In Toronto no television channel, including the CBC, has more than a 15 per cent share of the audience. The perceived legitimacy of public funding for the CBC has declined accordingly.

But there are two confusions here. First, audience share is different from audience reach. Even if BBC1 takes up barely a fifth of viewing time (that is, when its *share* has fallen to around 20 per cent), it is still likely to be used regularly by a large number of viewers (its *reach* will still be 80 per cent).[2] As *The Times* (7 July 1994) said in its editorial on the White Paper the BBC's task is to make sure it offers something for everyone each week. The legitimacy barrier will only be breached if sizeable numbers of people stop using the BBC altogether. That has not so far happened in satellite/cable households, where the BBC reaches nine in ten viewers each week.

The second confusion is to think of the licence fee as a subscription fee. Television ownership is near-universal, making the licence fee *de facto* a hypothecated tax, paid by individuals for the BBC to be available, whether or not they watch it. That is why it is wrong to argue that viewers should be allowed to opt out of the BBC if they no longer watch it. Just as parents of children in fee-paying schools are required to pay for state education and patients in private hospitals to fund the National Health Service, so those who opt out of the BBC still have a duty to pay for it. We recognize that there are serious problems with the licence fee: it is regressive and, as Wall and Bradshaw have shown, it is 'the cause of considerable injustice' (1994 pp. 1198–9). Moreover the government's ability to set its level makes the BBC vulnerable to political pressure. But, on balance, we agree with the National Heritage Select Committee's description of the licence fee (in a report commissioned to Touche Ross 1993): it remains the 'least worst' method of funding the BBC.

Quango or public service?

To argue that the BBC should remain in public ownership is not to reject the changes in broadcasting of the last decade or those of the next one. Nor is it to deny that liberalization could improve the functioning of the broadcasting market. The new media world will not be totally different from the current one. Consumers will continue to use broadcast channels. Both because these channels exhibit the technical characteristics of market failure and for the reasons of public policy stated here, the BBC should continue to exist as a publicly funded, not-for-profit, public service organization. However, we do not believe that it should do so in its present form.

Obvious practical problems of implementation remain. How, without the discipline of advertising or subscription, is the BBC to be accountable to users and to society as a whole? How is it to ensure that it serves the needs and desires of its real audience rather than those its own management attributes to the UK public? How is the BBC to balance potentially conflicting commercial and public service aims? How is the BBC to work efficiently and effectively?

Organizational change is needed if public service broadcasting is to meet the threats and opportunities posed by changed circumstances. We propose two changes: to the relationships between the BBC's constituent businesses, and to the role and selection of the Governors.

Reconciling public service and commercial enterprise

The BBC's pre-eminent status as the UK's principal public service broadcaster must be reconciled with its development as a major international player and stimulator of the UK's audio-visual sector. That represents an exacting challenge. But the commercial opportunity has never been larger: global and new media markets are opening new potential revenue streams for the BBC's rights and brand. The BBC must develop as an international provider of programmes and services and thus contribute to the growth of the UK's audio-visual industries. Its developing relationships with UK media companies, notably Pearson, are welcome. However, it is necessary to ensure that the public interest does not go by default in these arrangements and that the BBC's effectiveness is not undermined by the potential conflict between commercial and public service goals.

Until now the BBC has been organized as a vertically integrated, end-to-end programme producer, scheduler, and transmitter of television and radio services. That structure is already being challenged. The introduction of an independent programme producers' quota shows how the BBC's institutional arrangements can stimulate the audio-visual sector. The establishment of internal markets within the BBC via 'producer choice', i.e. the possibility for BBC producers to opt out of the 'in-house' production facilities, suggests that the BBC can use its own resources more effectively and contribute to the development of efficient service providers outside the Corporation. These changes have been painful (and the paper chase created by producer choice is clearly ripe for reform) but have strengthened the BBC.

The move away from traditional structures should be completed. The

BBC should be given a dual mandate: to continue to provide public service broadcasting in the UK and to increase the revenues it generates through international sales and services. This dual mandate can best be implemented by an internal disaggregation of the BBC. Confusion over conflicting missions may be alleviated by matching each goal to a single organization. However, we are convinced by the Peacock Committee's finding that the BBC must remain large so that it is better able to preserve its independence of government. We therefore recommend building on the reorganization of the BBC now being undertaken by the Director General and the Board of Management.

We propose the creation of arm's-length entities each of which would trade with other elements of the BBC and with service providers, clients and partners outside the BBC. For television (radio would follow a similar model) we propose to separate:

- scheduling and channel control
- programme production
- transmission and engineering
- commercial exploitation.

These four distinct, semi-autonomous entities would report to the Board of Management. The individual units would be given line management responsibility to minimize the paperwork and bureaucracy that have dogged the introduction of internal markets under producer choice.

The BBC would then more closely approximate to the federal business model of Unilever and ICI or, to take an example from the media, Chrysalis, than to the vertically integrated one of traditional nationalized industries. End-to-end integration may have been appropriate for a natural monopoly market with a unique funding method, but it creaks when the BBC is targeting a multitude of competitive markets and income sources.

Each unit would enjoy substantial autonomy to pursue its mandate. For example, BBC Channels and Programmes would focus on providing public service broadcasting in the UK, unencumbered by commercial imperatives. BBC Commercial would exploit commercial opportunities in all other relevant markets. The greater freedom should deliver greater efficiency and effectiveness throughout the BBC's activities. The Board of Management would be responsible for managing the relationship among the four units, whose respective mandates would be:

- *BBC Channels*: to deliver public service television across the range of media that would be available in the future (including satellite channels such as UK Gold, or video-on-demand ventures). A difficult issue remains over the relationship between subscription funded and free-to-air. But the overriding principle must be free access for all viewers and listeners to all BBC programmes (although a programme or service may be made available to paying customers earlier than it is to licence fee payers in the UK).

- *BBC Programmes*: to trade with BBC Channels, BBC Commercial and non-BBC broadcasters whether in co-ventures or as single clients.
- *Engineering and Transmission Services*: to compete against rivals such as BT, National Transcommunications and cable companies for the provision of distribution services both to the BBC and to non-BBC clients. BBC Engineering would be encouraged to seek commercial partners and profitable exploitation of its innovations.[3]
- *BBC Worldwide*: to exploit the BBC's expertise, rights and brand name in all areas other than the UK radio and television markets.

In their response to the White Paper, the Independent Television Commission suggested that the assets and rights that BBC Commercial would seek to exploit should be 'franchised for a period of years to a wholly commercial undertaking selected by competitive tender' (ITC 1994). This would meet three problems they foresee for an organization with both commercial and public service activities, namely:

- that the BBC 'is not in a position to balance successfully the mixed cultures of public service broadcasting and commercial business enterprise'
- that licence fee income might be 'put at risk by investment in commercial ventures'
- that the BBC has 'no risk or equity capital available to commit to new business ventures or partnership' (1994 pp. 1–2).

We share the ITC's concerns, but not its enthusiasm for franchising. In defining the franchise terms, an unsatisfactory choice would have to be made over whether to include the right to originate programmes and provide services under the BBC brand. Inclusion might undermine the brand and mislead consumers by allowing a non-BBC company to commission and make BBC programmes. Exclusion would prevent the BBC from exploiting opportunities that would benefit both its consumers and its finances (such as UK Gold or BBC Worldwide).

Instead, BBC Commercial should be given commercial freedom within the public sector. This suggestion would extend to the BBC's commercial activities the Post Office Ltd model developed by London Economics (1994b). BBC Commercial Ltd would be established as a fully fledged limited liability company, with the BBC the only voting shareholder. Its powers would be defined by its Articles of Association and a Statement of Corporate Conduct: these would bind the company only to undertake activities that were not being pursued by the public service units of the BBC, and prevent it from interfering in editorial decisions of the BBC's public service units. It would give the shareholder (in effect, the BBC Governors) the right to scrutiny and approval of acquisitions and investments. These safeguards would ensure that BBC Commercial Ltd did not undermine the BBC brand name or crowd out the BBC's public service activities.

This structure would make clear how commercial revenue should be shared between units. As a separate company, BBC Commercial Ltd would have to pay market prices for the assets and rights that it bought from other BBC units. This would ensure that the commercial value of the BBC's assets – i.e. the price paid for them by BBC Commercial Ltd – would flow back to the public service units, and hence to the licence fee payer.

These trading arrangements would have to be fair to competitors. In particular, the BBC should not be able to cross-subsidize BBC Commercial Ltd from public service funds and assets. The BBC's *Fair Trading Commitment* (1994b) that trading subsidiaries should produce and publish separate audited accounts should be followed. However, such public accounting will not necessarily demonstrate that the relationships between the separate arms of the BBC are conducted at arm's length. Accordingly, the Governors should have a special responsibility to ensure that the BBC's internal trading relationships are properly conducted and audited. Where there are grounds to believe that cross-subsidy is occurring between the public service units and BBC Commercial Ltd, and that the Governors have not been effective in ensuring proper compliance and audit, the matter could be referred to the Office of Fair Trading for review and, if necessary, to ensure redress.

As well as curtailing cross-subsidy, the creation of a limited company would expose the BBC to commercial freedoms and disciplines. Its financial freedoms would be those of any other limited company. There would be no reason for its finances to appear in the public sector accounts. BBC Commercial Ltd would be free to borrow or issue debt (or, if it chose to, offer non-voting equity) to finance new ventures and acquisitions. There would be no need for the Department of National Heritage to continue to set a limit on the BBC's borrowing, nor would the government be guaranteeing these loans against bankruptcy – so the BBC would not obtain finance at unfairly favourable terms. This possibility of bankruptcy, along with the requirement to provide a commercial rate of return to shareholders and creditors, would ensure that BBC Commercial Ltd felt the effects of commercial discipline as well as the benefits of commercial freedom.

The increased premium that the BBC's brand and programmes could secure in the multi-channel environment could top up licence fee revenue, as long as the government does not claw back new revenue by reducing the licence fee. Doing so would negate any incentive to commercial activity for BBC Commercial Ltd. These arrangements should free the BBC to exploit commercial opportunities, and thereby strengthen the BBC's finances to the benefit of UK viewers and listeners, but not to the detriment of the BBC's competitors.

Serving the public

Disaggregating the BBC would allow the non-commercial entities to concentrate on public service. The Governors would oversee their performance of that mandate and assure their independence from the aims of BBC Commercial. To do so, the Governors should be the interface between the public and the BBC, but currently the government is inserted between the two. It wields the public's power of selection and, as the Charter Review process has made clear, is the main body to whom the BBC is accountable.

Properly, there is great sensitivity about the relationship between public service broadcasters and government. In June 1994 there was a cautionary tale. The newly elected Prime Minister, Silvio Berlusconi, claimed that the Italian public service broadcaster, the RAI, was out of line in its criticism of his government. Other public service broadcasters, Berlusconi claimed, did not raise their voice against the majority which sustains the government. Berlusconi's claim was promptly rebutted by the President of the European Broadcasting Union, Albert Scharf (1994), who stated that 'public service broadcasting as it exists in Europe today is, by definition, independent of government and parliament.' This rebuttal did not stop Berlusconi removing the RAI's Governors and requiring the RAI to carry government funded advertisements for his economic reforms.

It is in the spirit of Scharf's definition that we propose changes to the role and selection of the BBC's Governors. With its Governors and funding chosen by government, the BBC risks being the country's largest 'quango'.[4] We cannot rely on the benevolence of our governments to guarantee the BBC's independence. The BBC's institutions must be reformed to pass the Berlusconi test: that is, would they survive a British equivalent of the former Italian Prime Minister?

How are quangos to be reformed in circumstances like this, where government control is not appropriate? We suggest as a guiding principle that power should be devolved both to users and, because of the BBC's social and political importance, to society and the polity as a whole.

Accountable to the public Broadcasting and broadcasters must be accountable:

- to ensure that the character and content of broadcasting services reflect the needs and desires of present and potential users
- to ensure that broadcasting performs its important social and political role
- to ensure that there are means for independent adjudication of and redress for users' complaints.

Our suggestions for improving accountability draw on those made by the consumer movement, whose involvement across the range of public services has grown sharply in recent years (Consumers' Association 1993). We support the proposal for a Broadcasting Consumer Council put forward by the Consumer Broadcasting Liaison Group, convened by Voice of the Listener and Viewer.

The White Paper rejected this proposal but did, however, agree to merge the Broadcasting Standards Council and the Broadcasting Complaints Commission. If it is to represent the consumer interest effectively, this new body must have powers to secure evidence from broadcasters and to require broadcasters to provide suitable redress in respect of complaints that have been upheld.

A wider strategy is also possible: creating a *one stop shop* to regulate all broadcasting independently from delivery mechanisms, or type of ownership. If a single regulator were to be established, the BBC should be brought under its remit.[5]

The role of the BBC Governors should also change. In 1993, the Governors themselves published *An Accountable BBC* (BBC 1993): a laudable attempt to clarify their role and streamline the structure of advisory boards within the BBC. But, amid sensible reforms, one fact stood out: the document is practically anonymous. It is signed by the Chairman and the Governors of the BBC, but their names do not feature anywhere.

Sir Kenneth Bloomfield, Lord Cocks, Mrs Janet Cohen, Dr Jane Glover, Lord Nicholas Gordon Lennox, Sir Graham Hills, Marmaduke Hussey,[6] Dr Gwyn Jones, Bill Jordan, Mrs Shahwar Sadeque, Mrs Margaret Spurr, Sir David Scholey: how many members of the public know that the common link between these mostly unfamiliar names is that they are all Governors of the BBC, and thus the trustees of the public interests? BBC Governors, as is widely recognized, are unrepresentative, unaccountable and endowed with no clear mandate. They are, at best, 'sound chaps' appointed by the Prime Minister, who can be relied upon to use their safe hands with discretion. At worst, they are placemen and women who have genuflected to political masters by censoring the

'Real Lives' series (BBC 1985), and ignored the wishes of viewers and listeners by rubber-stamping a new remit for Radio 5.

The BBC aims to be accountable to the public, but its mechanisms of accountability are deeply inward-looking. They need to be opened up to public scrutiny and influence. We suggest two clear principles for doing so:

- The Governors' role is to represent the public.
- The public should have formal and informal influence over the Governors' performance of that representative role.

Representing the public   The role of the Governors and their relationship to the management of the BBC has been contested since the inception of the BBC. Under Lord Reith, with the Director-General ascendant, the Governors were weak. Latterly, with a dominant Chair of Governors, the pendulum has swung the other way and the Governors have shown their power to fire and hire the Director-General of the BBC. Paradoxically, the government's White Paper *The Future of the BBC* vests effective power in the BBC's management by defining the Governors' role as one of approving rather than setting 'objectives for the BBC's services and programmes' (Department of National Heritage 1994 p. 3).

However, the role intended for the Governors has been clear from the start. The 1925 Crawford Committee meant the Board of Governors to be a supervisory board, of the type that evolved in continental businesses (Crawford Committee 1926). Although traditionally such dual structures were not favoured by British business, the corporate scandals of the 1980s have caused them to be championed again, most notably by the Cadbury Report on the *Financial Aspects of Corporate Governance. An Accountable BBC* (BBC 1993) is inspired by these trends in corporate governance, but does not fully translate that inspiration into implementation even though the Governors claim to have complied with the Cadbury Code (see BBC 1994b p. 63).

For the principles of the Cadbury Report to be truly extended to the BBC the first step should be to recognize that the Governors are essentially non-executive directors. They are not there to be regulators – they will always be too close to be impartial – although they do play a regulatory role, representing the interests of the BBC's various stakeholders. In general, as Labour's industry policy paper *Winning for Britain* (Labour Party 1994) argues, a representative role conflicts with a managerial role. To avoid that conflict of missions, we propose that the BBC be reorganized along the lines proposed in *Winning for Britain* with 'a two-tier board structure, in which a supervisory board of non-executive directors sets objectives and monitors the performance of an executive board with managerial freedom' (p. 1).

The Governors' role flows clearly from this definition: they should represent stakeholders by setting and monitoring the general direction of the BBC. The Board of Governors would be charged with establishing a rolling three year 'mandate' for the BBC, which specified performance indicators (PIs) in respect of the range of programmes, the economic performance of the Corporation, and the level of achievement required in respect of audience reach, share and satisfaction. These PIs would refer to the economic and public service roles of the BBC and, when they conflict, will define the balance to be struck between them. Thus far our proposals are readily compatible with established practice; the PIs set by the Governors are reported in the BBC *Report and Accounts* (BBC 1994a p. 15).

However, our proposals in respect of the relationship between Governors and BBC management, and Governors and the public, diverge from current practice. We propose that the responsibility for meeting these PI targets should rest with the Board of Management and with the Director-General in particular. Governors would have the power to secure whatever information they may reasonably require from the Corporation, but they would not have the power to view programmes before transmission. They would have the responsibility to support and encourage the broadcaster in the production and transmission of a proper proportion of innovative, iconoclastic and investigative programming.

To make sure the Governors carry out these duties in the interests of the public, they must be accountable to the public. In the past, formal methods of accountability were thought unnecessary. A culture of impartiality in public service was thought to be sufficient. Alternating parties in government would assure political balance. However, more than fifteen years of one party Conservative rule and its systematic attack on the culture of public service have undermined any confidence in the impartiality of quangos. It would not do for an incoming Labour government merely to replace the other side's placepeople with its own. We now need *informal* and *formal* methods of public influence to guarantee accountability.

Informal public influence   For the Governors to be accountable, they must be subject to open scrutiny, the public's *informal influence*. For scrutiny to be effective, relevant information needs to be made clear and available. First, the specific performance indicators set by the Governors should be made public, as should the BBC's success in meeting them. The raw data on audience behaviour should be in the public domain.

Some sectors of the industry have been sceptical about the reliability of these statistics, and about whether such monitoring would undermine the BBC's ability to take risks and fail. The proposals would have to be sensitively implemented to meet both these concerns.

Second, the Governors should make clear why they have reached particular decisions. Governors' decisions over 'Real Lives' and arrangements for the Director-General's remuneration, amongst others, have focused attention on the secrecy which clouds the deliberations of the Governors. The minutes of these and other more mundane meetings should be made public, after a suitable delay to allow for commercial confidence. Matters of commercial or national confidence, as well as sanctions, could be kept confidential if the House of Commons National Heritage Select Committee, to whom the Governors must report in such circumstances, so requires.

Formal public influence   The BBC's accountability could also be bolstered by *formal* mechanisms of public *influence*. To ensure that Governors are more representative of the public than of the government of the day, we propose devolving the power to appoint the Governors.

There is a developing consensus that power should be devolved to the most appropriate level. However, many of the ways of doing so in other areas (*Economist* 1994b) are not applicable to broadcasting. Direct elections work for some local bodies, but general suffrage of licence payers would be impractical and unlikely to generate meaningful turnout. Democratic control by elected councillors or MPs would be appropriate for some quangos, but clearly not for the BBC: not all politicians can be trusted to put the public interest above their vested interest in the political process. The challenge is to make broadcasting responsible to the public without being answerable to its rulers (Independent Broadcasting Authority 1974), *or* to other intermediary vested interests.

Devolving power

We therefore need to find new ways of devolving power. We propose:

- choosing the Governors through an open and public selection process, guided by pre-established criteria and performed by the relevant House of Commons Select Committee (currently the National Heritage Committee)
- replacing the current plethora of advisory committees with citizens' juries.

The selection process The aim of the selection process would be to devolve and separate power: *devolve* it to the public to increase responsiveness to users and buttress independence from government; *separate* it into different representative groups to reduce the risk of domination by any one group. The selection criteria should guide the process to achieve the following composition of the Board of Governors:

1 A regional component, so that all UK regions are represented in the Board. A number of posts must be earmarked for candidates of each region (or group of regions), covering a regional population of around 3 million people. However some regions (such as Wales and Northern Ireland) could be deliberately 'over-represented', because they are too important not to be formally represented despite population sizes below the threshold level.

2 A representation of special interest groups in broadcasting, such as children's needs, educational, arts and sports associations. It might also seek to go beyond this rather dated concept of representativeness to include individuals who can speak authoritatively for other social categories, such as age (younger and older people), ethnic minorities, people with disabilities and those who do not go out to paid work.

Governors would be appointed for a fixed term of office (between three and five years seems appropriate), would retire in rotation and would be eligible for one term of reappointment.

The Select Committee of the House of Commons would appoint the Governors of the BBC in public hearings, interviewing shortlisted candidates who have responded to the public notice of vacancy. Both the shortlisting and the selection criteria should be made public. The reasons behind the final selection should also be made public. The effectiveness of these arrangements would depend on the calibre of the Heritage Select Committee. This greater executive responsibility, and its authority over the release of commercially sensitive information, should encourage high calibre appointments to the Committee.

Introducing democracy is a leap into the unknown. So the new structures could be introduced gradually, initially perhaps selecting half of the Board in this way. Any weaknesses that emerged, perhaps in the calibre of Governors selected, could be addressed before extending the system to the whole Board of Governors.

The BBC was recently described as the second most important institution in the country. Such power cannot be wielded without accountability. Even if the system needed perfecting, the principle of election embodied in a Parliamentary Select Committee would be a clear improvement on direct patronage. It should separate the power of appointment from the government, make the BBC's Governors accountable to the public and diffuse power so as to avoid domination by any vested interests.

Citizens' juries Once chosen by the electoral college, the BBC's Governors could keep in touch with – and be guided by – the views of the public through the use of citizens' juries, held at regular intervals in each region of the country. These are small groups of people chosen at random from the electoral register who meet together to discuss specific issues and draw conclusions. They hear evidence from witnesses and cross-examine them, and have time for a full discussion of the issues at stake. This model of decision making has been tested in Germany and the United States and is examined in detail in a report from IPPR (Stewart, Kendall and Coote 1994). Citizens' juries would allow the BBC to be more responsive to the demands and tastes of the public at large. Unlike consumer groups, citizens' juries do not represent particular interests but are made up of ordinary citizens with no special axe to grind. Unlike opinion polls or surveys, they provide an opportunity for deliberation and informed decision making.

   Citizens' juries could be consulted about the performance indicators the Governors set. But they could also sharpen debate on particular questions, such as degrees of violence and sexual explicitness on television, programme mix, whether court proceedings should be televised, privacy and other ethical dilemmas. Juries' decisions would guide but not bind Governors.

Summary

1 Public service broadcasting is being challenged across the world. New technologies promise narrowcasting and overcoming of market failures. Governments are eager to privatize public sector institutions.
2 Rejecting privatization of the BBC, we argue that a reformed BBC can find a role for this new era.
3 Reform should move along two principal lines: as the country's largest quango, the BBC must be made accountable; as a public service with commercial goals, it must be given commercial freedoms, but be prevented from cross-subsidizing to the detriment of competitors. Specific proposals include:
   (a) electing the Governors and creating mechanisms of informal public influence – publicity of performance indicators and motiva-

tions to specific decisions – to make the BBC more responsive to the public

(b) disaggregating the BBC into autonomous units with public service and commercial mandates – respectively, scheduling/channel control and programme production, and transmission/engineering and commercial exploitation – to improve efficiency and remove conflicts of interest

(c) creating a BBC Commercial Ltd, owned by the BBC, with the freedom to borrow without inclusion in the PSBR and the ability to sell non-voting equity

(d) maintaining licence fee funding and services that are free at the point of use, while restricting charging – via subscription or advertising – to incremental channels.

8 Convergence and Change: Reforming the Regulators

Media and communications law and regulation cry out for reform. Changed circumstances make established institutions and instruments no longer effective. Too many bodies confuse consumers. Remits formulated for distinct media are poorly adapted for the contemporary changing markets and wastefully overlap. Decisions are taken by unrepresentative and unaccountable gatekeepers who are expected neither to make decisions openly nor to give public account for their decisions.

The time honoured UK 'strategy of having no strategy' towards media policy, to use the words of James Curran (1995), might have been tolerable in the past, but is now impossible to accept. Yet the sense of uncertainty aroused by rapid change means that a conservative attitude continues to prevail. Regulatory arrangements need to be promptly adapted to technological change in order to bring closer to realization the vision of citizens' access to information and media independent of dominant power centres. Pioneering such a regime will better realize the entitlements of UK citizens and consumers and serve the economic interests of UK media and communications industries. The UK has been a regulation laboratory which other jurisdictions have taken as exemplary: if this demonstration effect continues UK firms will be familiar with regulatory regimes adopted elsewhere. The UK's unique advantages make this an attainable vision – but one which requires clear thinking about the social, economic and political goals of media and communication policy and a re-evaluation of the UK's regulatory institutions.

The regulators

There are no less than ten statutory and self-regulatory bodies for media and communications in the UK, and arguably even more. Broadcasting and telecommunications are regulated by statutory bodies. The Independent Television Commission (ITC) licenses and oversees cable, satellite and terrestrial commercial television. The Radio Authority (RA) does the same for commercial radio. The Welsh Fourth Channel Authority runs the largely Welsh language television service S4C. The Broadcasting Complaints Commission (BCC) is best known as a 'poor man's libel court' and adjudicates on complaints of misrepresentation on radio and television. The Broadcasting Standards Council (BSC) adjudicates on complaints concerning breaches of proper standards of taste and decency in broadcasting. Telecommunications is regulated by Oftel, established by statute under the Telecommunications Act 1982.

Four self-regulatory bodies are established by the industries which they regulate and work with varying degrees of closeness with the statutory regulatory bodies. They are the Advertising Standards Authority (ASA), the British Board of Film Classification (BBFC), the Independent Committee for the Supervision of Standards of Telephone Information Services (ICSTIS) and the Press Complaints Commission (PCC). Curiously the BBFC, though a self-regulatory body, has statutory responsibility for video classification and its film classifications are becoming increasingly important in television broadcasting regulation.

Other bodies discharge other regulatory functions. The BBC is outside the remit of the Independent Television Commission and the Radio Authority (but not the BCC and BSC) and is regulated by its own Governors who are established under Royal Charter and are for legal purposes themselves the BBC. The Radiocommunications Agency allocates the radio frequency spectrum to both broadcasting and telecommunications. The Monopolies and Mergers Commission and the Office of Fair Trading exercise jurisdiction over all the media and communications industries. The Monopolies and Mergers Commission has special responsibilities for newspaper mergers under the Fair Trading Act 1973. Moreover, the Department of National Heritage, the Department of Trade and Industry, the Council of Europe, the European Commission, and the International Telecommunications Union also shape the environment for media and communications in the UK. In so far as they promulgate rules and negotiate binding international agreements, they could also be regarded as regulators.

Appendix 2 gives details of the membership of the UK regulatory bodies, which are summarized in table 9 for convenience.

Table 9 Overview of UK regulatory bodies

Name	Function	Status	Average age	Male–female split	Number of non-white members
Broadcasting Standards Council (BSC)	Advises on matters of taste and decency, considers complaints and commissions research into public attitudes	Statutory	58	4M–4F	0
Advertising Standards Authority (ASA)	Promotes and enforces standards in all non-broadcast advertisements	Self-regulatory	n/a	7M–6F	1
Broadcasting Complaints Commission (BCC)	Deals with complaints about unfair treatment in TV programmes and infringement of privacy	Statutory	60	3M–2F	0
Radio Authority	Licenses and regulates independent radio services	Statutory	50	4M–3F	1
Welsh Fourth Channel Authority	Regulates the Welsh public broadcaster S4C	Statutory	n/a	4M–3F	0
Press Complaints Commission (PCC)	Creates and enforces codes of practice, informal resolution of complaints	Self-regulatory	60	10M–3F	0

Independent Television Commission (ITC)	Licenses commercial TV services; regulates to promote choice and quality; defines quality	Statutory	56	6M–4F	1
Independent Committee for the Supervision of Standards of Telephone Information Services (ICSTIS)	Supervises promotion material and content of premium rate telephone services; handles complaints	Self-regulatory	est. 46	4M–6F	1
BBC Governors	Regulate the BBC	Statutory	59	7M–2F[a]	0
Oftel	Regulates UK telecommunications licensees; promotes competition and protects consumer interest	Statutory	n/a	1M	0
British Board of Film Classification (BBFC)	Responsible for all matters relating to the classification of film and video works	Self-regulatory	65	3M–2F	0

Source: IPPR Research 1995

[a] Three positions were vacant.

Why regulate?

The benefits conferred by technological change and competition in UK media and communications are striking. New broadcasting services like satellite television and commercial national radio have provided new consumption opportunities and encouraged incumbents to change their programming. The price of telecommunication service has fallen and its quality has risen dramatically. Increased competition, however, has not stopped big firms, public and private, from dominating media and communication markets. In spite of enormous growth in the number of licensed telecommunications operators in the UK, BT is still strikingly dominant with 92 per cent of UK telephone lines and 87 per cent of the value of business. The BBC has 19.7 per cent of the share of voice of the UK media market. News International has four national newspapers and the largest shareholding in BSkyB, the dominant satellite television. Some programming – like sports – hitherto available to viewers at zero cost, is increasingly available only to those who pay. The goal of universal service at affordable cost is under threat in broadcasting and telecoms. The potential control of markets afforded by ownership and control of key 'bottlenecks' shows that media markets cannot be left to themselves.

Regulation is not just concerned with monitoring dominant firms, granting access to bottleneck facilities and correcting market failures. It has a positive duty to promote the public interest, that is to pursue policies which promote the welfare of UK citizens, firms, and consumers. In the introduction to this book we have presented the criteria that should guide public policy in the communications field. The next section analyses how far existing policies have succeeded in fulfilling them.

Auditing the UK system

Security, opportunity, democracy and fairness are the criteria for policy making and evaluation elaborated by the IPPR Commission on Social Justice (IPPR 1993) which we have applied to media and communications. How does the current UK system measure up to such criteria?

The first policy objective, *security*, which demands relief from the fear of poverty, is guaranteed by accessible information and means of communications. Universal service at affordable price in telecommunications and broadcasting is one of the most important policies responding to this criterion. Here, the UK record is creditable. Quality newspapers, priced between 30p and 70p per issue, are affordable by all except the very poorest. Penetration of voice telephony to 93 per cent of homes is a notable achievement, but one which can still be improved. All UK communities of more than 200 people have access to four channels of terrestrial television for an annual licence fee of £90. However, some recent policies seem to move away from improving cheap universal access. Revisions to European intellectual property law have lengthened the term of copyright protection. Harmonization of EU copyright law has subordinated the 'right to copy', implied in the word 'copyright', to the interests of owners of intellectual property. Copyright law should strike a balance between authors' rights to remuneration and the public right to copy. It needs to protect the interests of producers in securing a fair return for continuing production of new works, as well as the interests of consumers in cheap and easy access to information. But protection has now been extended – seventy years *after* the death of the author – far beyond what is required to stimulate production of new works. Not only will this impede formation of a well informed polity, able to decide and deliberate rationally, but copyright owners will be able to control information bottlenecks at the expense of knowledge and the useful arts.

Policies promoting *opportunity*, the second criterion, are designed to improve life chances. In turn this goal is ensured by two parallel policies: promoting widespread use of information and means of communications and maintaining high standards of quality. Judged by this wider definition, the record of UK media and communications is mixed. The use of media and communications is near universal, and UK television and press enjoy an international reputation for high journalistic standards. However, there is still a significant minority of untelephoned homes (7 per cent of UK households) which therefore have no access to emergency services or family and friends in moments of crisis. Also, there are striking gaps in the UK media's coverage. For example, nowhere is there comprehensive and authoritative coverage of the activities of the European Parliament and Commission.

A second important way to promote opportunities is the design of rules to improve choice. To assess the extent to which there is real choice in UK media and communications services is difficult. A consistent motif in discussion of UK media, particularly of the press, is deep alarm at a growth in concentration of ownership and perceived bias against political views – particularly left views – outside those sanctified by the whips' offices of the major parties.[1] But newspaper buyers can choose between some ten national papers of which at least five, judged by any reasonable international standard, provide comprehensive and well founded accounts of events and issues. Left and right commentators outside the Parliamentary consensus have consistently argued that their points of view have been systematically excluded, misrepresented or marginalized. Thus far, government has not intervened in the press, sharing with the opposition parties the belief that the cure of intervention is worse than the disease of a partisan press.

In telecommunications choice among services and carriers has undoubtedly grown over the last decade but new services have largely been provided just for business users. In broadcasting, choice was enshrined as an official policy goal in the 1988 White Paper on broadcasting (Home Office 1988). Many new channels have extended choice. However, growth in choice between channels may not enhance choice between types of programmes: the very existence of programmes from which returns are low relative to costs may be threatened. Programmes targeted at the poor and old may be cases in point.[2] Taking the media together, UK consumers do have access to a range of viewpoints and analyses. But regulation can extend both the range and the representativeness of the viewpoints voiced in the media, and improve the scope and availability of telecommunication services.

Opportunities are increased by the ability to make informed choices. Information about programmes and services has been improved, thanks to the requirement that service providers must license other firms to publish their programme schedules. But more can be done to improve consumers' ability to choose well. For example, much broadcasting audience research remains confidential, as do the grounds on which important regulatory decisions are taken (such as the allocation of franchises). Details of the pay of the Director-General of the BBC were not even disclosed to members of the Board of Governors by the Chairman before public scandal forced their disclosure! Information in the public domain is not sufficient for the legitimate purposes of either media users or regulators.

Democracy, the third criterion, means equal participation in decision making by members of a community. Neither markets nor democracy work if participants in economic and political life lack information. This criterion points strongly to laws supporting freedom of information and freedom of expression. Second, it points to representative and accountable governance of communications and media. Democracy requires the exercise of control through a strong civil society rather than a more powerful state in order to prevent the exploitation of power – whether in market or political domains. Here, UK media and communications are remarkably deficient. Public information is scarce and expensive. Freedom of expression is nowhere guaranteed. Regulators are not accountable to citizens and concern about the concentration of corporate power in media and communications is pervasive. Formal representation of user views, needs and interests is poorly developed.

In broadcasting and telecommunications, representation is confined to advisory committees appointed by the regulators and broadcasters rather than by consumer bodies or by users themselves. Representation does not exist for the press. The first ever public hearing held by a UK regulator took place in November 1995. Oftel is to be congratulated for its welcome innovation.

Finally, the fourth objective, *fairness*, points to policies aimed at enhancing respect for the rights and differences of minorities, to redress of injustice and protection of the vulnerable. Redress is difficult, at best, to obtain. As the Consumers' Association stated, the crowd of regulators means that 'there is a potential for confusion over which body to approach and over the effect that complaining and regulation will have' (1991 p. 32). Recourse to the courts in cases of alleged libel is a rich person's prerogative. The 'poor man's libel court', the Broadcasting Complaints Commission, is slow; it judges many of the complaints it receives to be beyond its jurisdiction. In any case, the BCC provides redress only in broadcasting. The McWhirter[3] case established that television viewers had no standing to challenge the decisions of those who controlled the system or to call them to account.

The Press Complaints Commission (PCC) offers redress for offensive behaviour by the press. There are recurrent complaints of misrepresentation and invasion of privacy by the media, particularly by newspapers. Its predecessors, the Press Complaints Commission Mark One (1990–3) and the Press Council (1953–90), were severely criticized, notably by Sir David Calcutt (1993). Although the new PCC has made striking improvements in its complaints handling, grounds for concern remain. If newspapers and magazines do not acknowledge the PCC's authority, or refuse to adhere to its judgments,[4] complainants have no redress other than through recourse to law. PCC sanctions are weak: the requirement to publish an adverse adjudication is not a convincing punishment, or a deterrent to offensive behaviour. The UK has no tort of privacy. Moreover, the very success of the PCC in securing satisfactory informal resolution of the majority of complaints means that the possible deterrent and demonstration effect of its judgments is weakened. Finally, the PCC lacks legitimacy because of the mode of appointment of its commissioners and its financing by the industry which it regulates.[5] In sum, fairness demands change: the interests of UK citizens and consumers could be better served by new regulatory policy and practice.

Sector-specific regulation?

Is there still a need for sector-specific regulation of media and communications? Markets do not always work and when they do not regulation is required. But the character of markets changes. Failed markets may no longer fail, well functioning markets may stop operating well, and separate markets may converge. We can no longer take for granted that media are separate – that the press is a domain separate from broadcasting and that broadcasting is distinct from telecommunications. Nor can we assume that broadcasting and telecommunications markets will always fail.

Latterly some have claimed that changed circumstances mean that many goals of industry-specific regulation could be achieved through 'generic' regulation – based on new competition laws and specific statutes on such issues as privacy and freedom of expression. Leo Gray, an Australian media lawyer formerly with the Australian Broadcasting Tribunal, has argued for a self-regulating media system based on competitive markets and policed, like other markets, by a competition authority: 'There is very little point in having any detailed structural rules for regulating ownership and control ... the only objective that seems capable of being realistically pursued is the encouragement of real competition' (1992 p. 22).

Though we favour competition, and a stronger UK competition regime, we are not persuaded that competition is likely to make sector-specific media and communication regulation redundant. Technology has reduced the degree of market failures, but both media and communications are still failed markets and neither freedom of expression nor effective competition can be secured without a framework of law to define and enforce rights. The shaping of media and communications to achieve social goals must be done by agencies independent of government. Political authorities might rig markets to favour clients or to reduce the freedom of the press and other media, a freedom vital to democracy. Moreover regulators must have specialized knowledge and skilled judgement to regulate well. Day-to-day regulation bears on complex situations which cannot adequately be encompassed by the application of simple prescriptive rules.

Although it is hard to imagine circumstances in which sector-specific media and communications regulation can disappear, there is certainly room to consider whether a different balance between sector-specific regulation and a general competition policy would be beneficial. We support those who have argued for a stronger UK competition policy,[6] believing that although this cannot do everything it can do a lot. We also agree that the overall health of the media and communication sector will be strengthened if subject to 'prohibition' competition policy, that is to say a policy which generally prohibits behaviour with anti-competitive intention or effect.

In 1992 the government published a Green Paper *Abuse of Market Power* (Department of Trade and Industry 1992) aimed at reforming competition law. It subsequently favoured the least demanding of the three options canvassed in the document, which proposed only minor changes to the *status quo*. Instead, we support the conclusions of the National Consumer Council that 'simplification of the UK law, and the closer alignment of our rules with those of the European Union, would be far more constructive than tinkering with the existing law', not least because oligopolies 'remain a prominent feature of the UK economy' (1995 p. 70).

The principles of EU regulations provide a valuable basis for strengthening UK competition policy and regulation: strong investigatory powers for the competition authority, enforced by fines and empowered to reverse offensive actions. A prohibitory regime along established Community lines, but with suitably modified thresholds, would improve matters within the UK and bring the UK's competition laws into line with its European partners and important trading partners.

Competition policy is necessary to guard against undesirable concentrations of power and to permit regulators to intervene quickly when a dominant firm acts anti-competitively, a consideration stressed both by the Director-General of Telecommunications at Oftel's public hearing on 23 November 1995 and Professor Melody at IPPR's Seminar Law and Regulatory issues on the Superhighway in September 1995 (Collins 1996). But it is not sufficient to deliver a democratic marketplace for ideas or to provide a comprehensive basis for regulation of media markets. Strengthened UK competition law promises a welcome increase in powers to countervail abuse of market power but lacks the capacity either to redress market failure proactively or to secure important non-economic policy goals. Competition policy is not sufficient to establish a robust civil society animated by vigorous debate. Nor will it establish easy access for citizens to the information necessary to fully participate in political and social life. These objectives require proactive policies if they are to be realized. If all are entitled to the communication means for full participation in social and political life at reasonable cost to the community as a whole, regulation must be directed towards realization of such entitlements and universal service in a broader sense.

We can define universal service in media as an entitlement to 'expression', access to a 'voice' for diverse views. The UK has sought to secure this broadly defined universal media service and expression of a range of voices through the medium of broadcasting. It has devised two means for achievement of its goals: public service broadcasters (the BBC and Channel 4) and the requirement that commercial broadcasters advance these goals as a condition of licence. It has funded this project:

- through a hypothecated tax, the licence fee
- by granting monopolies (to sell terrestrial television advertising) and

requiring specific performance of these and other objectives, in exchange for monopoly privileges

- by allocating desirable radio frequencies to service providers charged with the achievement of such goals.

Curran (1995) has argued that the 'positive freedom' principle of universal media service, hitherto realized in the UK only through broadcasting, should also be realized through the newspaper and periodical press. He advocated establishment of a Media Enterprise Board to assist in the realization of this goal. However, in our view, evidence of market failure in the print sector is less striking than in sectors such as broadcasting and telecommunications. Rather than extending the universal service doctrine to the press, it is more appropriate to extend it to the infant industry domain of interactive telecommunication services – to the Internet and World Wide Web. State intervention in broadcasting stimulated the penetration of services and the habit of consumption. In the same way, the availability of high quality, not-for-profit public information, education and entertainment could stimulate usage of Internet services. Indeed public broadcasters in the United States, notably WGBH in Boston, are doing just that.

However the sustainability of the funding methods used to secure positive freedoms in broadcasting is likely to decline. National broadcasting regulation is no longer able to guarantee exclusive rights to the sale of television advertising in particular markets. Moreover, technological change has opened up more radio frequencies suitable for broadcasting – a process which is likely to continue. Trading off programme characteristics against the right to use the radio frequency spectrum and enjoy an advertising monopoly is a bargain governments may no longer be able to enforce. And few governments are likely to raise taxes to sustain increasingly expensive public broadcasting services – still less a subsidized press sector.

Maintaining state funding to secure positive freedoms in media and communications is a matter of a simple political choice and one we think is important for the government to take. Funding for efficient public services in media and communications could well come from revenue generated by the auctioning of the radio frequency spectrum.

Reforming the regulators

Regulatory innovation has gone hand in hand with the development of media and communications in the UK. The BBC was one of the first public sector industries and the ITA (Independent Television Authority, predecessor of the IBA and the ITC) has been claimed as the first US-style regulatory agency to be established in the UK (Baldwin 1995 p. 14). The Telecommunications Act 1984 introduced price cap regulation, which was echoed in UK gas, water, electricity and airport regulation.

Convergence between old media and the development of new media like the Internet gives rise to new problems of demarcation between the private and the public sectors. Telephone conversations are recognized to be private communications. Monitoring them requires specific authorization by the Home Secretary. Radio and television broadcasting is recognized to be public because no one (before the introduction of conditional access systems) could be excluded from access to them. But latterly encrypted broadcasts, which cannot be involuntarily consumed, have made broadcasting a more private medium. On the other hand, World Wide Web sites accessed via the Internet and dial-up collective chat lines have given telecommunications a quasi-public character.

Important questions of protection of the private experience of individuals from unwarranted intrusion by the media remain unresolved. The Williams Report (Home Office 1979) recommended restriction of access to potentially offensive material so that individuals were not exposed to involuntary intrusion of pornography into their lives, and so that their privacy was not invaded by material they found offensive. The boundaries between the public and private and authoritative adjudication on the offensiveness of particular messages remain matters of concern. The Calcutt Reports (1990; 1993) testify to the continued absence of consensus on what is legitimately a matter of private concern, and should not therefore be reported, and should be public.

The government charged the Calcutt Committee to consider the measures 'needed to give further protection to individual privacy from the activities of the press' (1990 p. 1 para. 1.1). Calcutt proposed strengthening protection of privacy by enactment of a privacy law and providing public access to redress through statutory handling of complaints. As Curran stated, this

> is an issue on which opinion is understandably divided. The danger of such a law is that it would tend to restrict critical scrutiny of the powerful ... It is also not clear that politicians, who have been on the receiving end of a critical press assault on their personal morality and professional probity on a scale unprecedented in this century, would in fact draw the balance between privacy and public interest in an acceptably libertarian way. (1995 p. 24)

The difficulties to which Curran points are real. On balance we take the view that statutory strengthening of the public's entitlements to privacy is likely to be in the public interest – provided that protection of individuals' privacy does not abridge the public's right to know about public matters. Inevitably discrimination between the private and the public is judgemental. There is therefore merit in enunciating the criteria for making such delicate distinctions after a process of public deliberation. Day-to-day adjudication between legitimate and illegitimate reporting of private matters (and some private matters of public figures are of legitimate public concern) should be undertaken by accountable and representative regulators. Neither condition is satisfied under present arrangements.

The gatekeepers who control UK media content are neither representative of the public nor accountable to them. The plethora of authorities with overlapping and contradictory responsibilities is inefficient. The resulting unequal application of regulatory authority and jurisdiction in different sectors is unfair. These arrangements speak of feudal muddle, patronage and preferment rather than what is appropriate to a modern state and to a vital sector of the UK economy. Research for IPPR, reported in appendix 2, shows that eight out of ten chairs of UK media content regulatory bodies[7] are men. Only one chair is younger than 60 (Baroness Dean). Five chairs are Oxbridge educated, and less than 5 per cent of the members of the regulatory bodies are non-white. Membership of the bodies is London dominated and all bodies are located in London. In contrast, the regulation of telecommunications is vested in one person, the Director-General of Telecommunications. Although we believe that the UK has been extremely fortunate in the quality and performance of its Directors-General of Telecommunications, we agree with Lord Borrie's judgment (at IPPR's Seminar on Law and Regulatory issues on the Superhighway in September 1995: Collins 1996) that a small college of regulators, like the United States Federal Communications Commission, is preferable to a single office holder.

Public sector suppliers have a significant impact on the quality and character of competing commercial offerings. Indeed Garnham has claimed that 'licence-financed public service broadcasting has done a better job of mimicking the perfect market than has advertising-financed broadcasting' (1993 p. 153). Channel 4, S4C, the BBC and Kingston Communications are significant in providing efficient television and telecommunication services and establishing a threshold of performance for commercial competitors. Although few public sector institutions remain in media and communications, public sector governance exemplifies the problem of accountability and representativeness evident in UK regulation. In fact, public sector governance is a matter of public concern. The problem is made starkly apparent when the remuneration of the Chief Executive of Channel 4 – an organization with an annual turnover around £400 million and 564 employees – is compared to that of the Director-General of the BBC, responsible for an organization employing 22,135 people with a turnover exceeding £2 billion. Why is the Chief Executive of Channel 4 paid 60 per cent more than the Director-General of the BBC? The performance of senior public sector executives is rewarded without any obvious link to output prices, consumer satisfaction or benefit.

How can we improve representativeness and accountability of regulators and the public sector? An obvious remedy is an open and public selection of regulators and governors. This could make gatekeepers accountable and provide a better channel for citizen and consumer concerns. It would fill the gap which now yawns uncomfortably between general concern about the concentration of proprietary power and the fear that state intervention may substitute a cure worse than the disease. This fear is a consequence of the remarkable centralization of the UK state, which makes pluralization and democratization of media power impossible without radical institutional change. In the preceding chapter we propose new means to select the BBC's Governors (formal influence via public selection process and citizens' juries and informal influence via publicity of information and decision processes). Such means would be appropriate for appointing the members of Ofcom.

Content regulation

Regulating carriage matters in the public interest is conceptually relatively straightforward, although implementation is likely to require increasingly specialized expertise in the regulator. The policies needed to promote the public interest are universal service at affordable cost; application of the 'any to any' principle; competition between infrastructure and service providers; and regulation where competition is imperfect.

The aims and conduct of content regulation are more difficult to specify. The difficulties of basing it on a licensing regime, like the ITC's, are increasingly apparent. The ITC is perceived to regulate satellite and terrestrial television services unequally. Effectively, it does not have equivalent jurisdiction over different transmission media. The disparity foreshadowed in the terrestrial/satellite case is likely to grow with new media. The ITC's claim to jurisdiction over images accessed via the Internet and transmitted over wired telecommunications is unsustainable because the ITC is unable to monitor such services and impose compliance to its requirements. Service providers using different delivery technologies must be treated even-handedly.

In chapter 5 we have discussed the regulation of offensive material and have proposed more liberal practices than now prevail: a 'right to know' Act, no prior restraint of publication of text, except in cases of clear and present danger, and restricted access to pornographic images.

Yet, giving priority to the principle of freedom of expression and calling for no prior restraint still requires judgement for suppression of offensive material *post hoc*, in some instances. Since distinctions between artistic and pornographic representations, or between factual reporting and sadistic voyeurism, are so difficult to define (and because there is a pervasive social demand that lines be drawn), appropriate institutions for making such skilled judgements are required. Content regulation should be undertaken by a representative body rather than through the courts.

John Trevelyan (1973 p. 160], the Secretary of the British Board of Film Classification (BBFC) for thirteen years, recounts how the version of *Psycho* circulated in the UK was cut by the BBFC. His account unveils the troubling appointments procedures of the Board he served: decision making was a private affair; examiners (censors) were cautiously selected from those thought to be sound. The Home Secretary was informally consulted about appointments, and processes of decision making skewed towards the interests of large industry players. Senior management of the BBFC was able to select and discharge examiners at will. Frank Panford, a former BBFC examiner, expressed concern about the powers enjoyed by the Chief Executive of the BBFC, who 'in effect sacked, called it redundancy, the whole body of the examining staff at the BBFC' (verbatim from IPPR's Seminar on Expression and Censorship, June 1995). This demonstrates the unsatisfactory and arbitrary character of censorship in Britain.

With the troubling exception of the BBFC, statutory regulation is undertaken by bodies appointed by government and therefore, in some sense, accountable to Parliament. But many mysterious and unaccountable guardians decide for us in the domains of the press, advertising, and telecommunication services. Self-regulatory bodies, which often elaborate successful content codes or have striven to provide effective handling of complaints, cannot claim to provide fully satisfactory representation and redress – still less access and information. Self-regulatory bodies are accountable not to citizens or consumers – even through the imperfect channel of Parliament – but only to the industry which has established them.

Decision making on such sensitive matters is so important for freedom of expression and for social order, that the composition of bodies regulating content and the character of their consultative and decision making processes are of capital importance. It is important that such decisions be taken by regulators:

- who are representative of the population as a whole
- who consult the public
- who are familiar with the changing texture of public values
- who are accountable to the public.

Different standards should be applied to different media and different types of material.

How many regulators?

Despite the content and carriage separation, consumers as well as experts are uncertain where jurisdiction lies. Overlapping and mutually blocking jurisdictions cause unsatisfactory decision making and inhibit intelligent assessment of risk and reward. Uncertainty about the basis for decisions and technologies falling outside the remit of existing regulators harms the public interest. As the Director-General of Fair Trading stated, the 'proliferation of regulatory bodies and regimes . . . requires some kind of rationalisation' (1986 p. 11).

UK media regulation has developed through a process of historical accretion. New bodies have been created for new regulatory functions but none has died. Institutions are periodically reshuffled: the Independent Broadcasting Authority was divided into the Radio Authority and the Independent Television Commission. The new ITC includes the previously autonomous Cable Authority. A reduction in the number of bodies is unusual.[8] Incremental change can work, if the regulated industries also change incrementally. At present a single firm can be regulated by several regulators, and firms competing in the same market may be regulated by different regulators. For example, satellite movie channels claim to satisfy their regulatory requirements when broadcasting films classified by the BBFC, whereas terrestrial television channels are required to satisfy the ITC's stricter criteria. As the Director-General of Telecommunications argued at IPPR's Seminar on Law and Regulatory Issues on the Superhighway (Collins 1996), regulation should be based on markets rather than technologies. Media markets are converging; so should regulation.[9]

Not all agree that it should. Media and communications industries insist on the merits of industry-specific regulation. Established regulators point to the specificity of their expertise and the fruitfulness of debate between themselves. But a plurality of regulators leads to fissures, anomalies and contradictions through which offensive practices can sometimes slip comfortably.

On balance we find arguments for fewer regulators compelling, and argue for a single regulator in the communications sector. Media and communications are becoming increasingly interconnected and it is less troublesome to conceive of them as an interconnected whole than as a series of discrete phenomena. Although content regulation and the regulation of markets are different tasks, they are not necessarily best done by separate bodies. Indeed the ITC and Radio Authority now regulate both structure and content. Each regulatory task is more efficiently discharged because both are done by the same body. The leverage which accrues from structural and carriage regulation assists enforcement of content regulation. Carriage and structural regulatory decisions, such as those concerning diversity of outlets, are influenced by considerations of content. To regulate concentration of ownership effectively, both the individual and the aggregate UK media and communication markets must be considered together. Consumers will be better served by one stop regulatory shopping. In 1992–3, 88 per cent of complaints to the BCC and 33 per cent of complaints to the BSC lay outside the terms of reference, and there are similar grounds for concern in respect of the ITC (Mitchell 1994). As the National Consumer Council stated: 'there may be a case for a single regulator of broadcasting ... there are compelling reasons for a single body to represent the viewer and listener interest and to handle complaints' (1994 p. 7).[10]

What is required is a single regulator – we suggest 'Ofcom' – applying common content guidelines and codes with different standards of severity and exactitude for different media and types of communication. For communications that can be consumed involuntarily, i.e. public media (posters on street, advertising sites and free-to-air radio and television), content codes will apply with greater severity than with media that are unlikely to be consumed involuntarily. Less severe content codes should apply to periodicals from the top shelf of a newsagent or to a licensed sex shop, to World Wide Web sites clearly labelled as a site where sexual material is to be found, or to a clearly labelled sex chat line. Private communication should not be subject to content regulation whereas public communication may be. The more private and voluntary the communication the less statutory content regulation should be applied. The more public and less voluntary the communication the stronger the case for content regulation.

Adjudication on complaints of misrepresentation and invasion of privacy should fall within the remit of Ofcom. Press regulation should be performed with a notably lighter touch than regulation of non-print media. Ofcom, moreover, will have some responsibilities for the structural, pro-competition regulation of the press – issues such as concentration of ownership. However, Ofcom shall have no powers of prior restraint over publication of the written word. Its powers to regulate content are to provide *post hoc* redress for those whose complaints it upholds and to reduce offences by formulation of codes of conduct. These should be agreed in consultation with interested parties, not least the journalists, editors and proprietors over whom its regulatory authority will be exercised.

There are significant advantages in subsuming the functions presently discharged by Oftel, the ITC, the BSC, the Radio Authority, the BCC and the statutory responsibilities of the BBFC in a single regulatory body. There would be further advantages if this body, Ofcom, also oversaw allocation of the radio frequency spectrum and undertook the regulatory functions presently discharged by the Governors of the BBC, who now have the unenviable task of being judge and jury in their own case.

Concentration of regulatory functions in a single agency would offer economies and help to ensure that Ofcom was sufficiently well resourced to do its job effectively. Here the ITC funding method, whereby the regulator is funded by the firms it regulates in proportion to their size and ability to pay, offers a promising model.

Re-regulation not de-regulation

Regulation should not wither away, but it must change. Two kinds of re-regulation are now called for: technology neutral law and a changed relationship between government and regulator. Canada provides a good example of the move towards technology neutral legislation. The Canadian Broadcasting Act 1991 defines broadcasting as 'any transmission of programmes, whether or not encrypted, by radio waves or other means of telecommunication for reception by the public by means of broadcasting receiving apparatus' (Section 1.2.1).[11]

Adoption of the new Canadian formula in the UK would reduce anomalies and bring new services such as cable broadcasting and dial-up video services within the remit of broadcasting regulation. It would reduce uncertainty, which inhibits innovation, and ease a transition from the current many agencies to a single regulatory agency like the Canadian Radio-television and Telecommunications Commission (CRTC) in Canada and the Federal Communications Commission (FCC) in the USA.

Technology neutral law implies a return to first principles and to what one Australian broadcasting regulator has called 'fuzzy law' (Brooks 1992 p. 9). Fuzzy law increases regulatory discretion. Australia has given its broadcasting regulator, the Australian Broadcasting Authority, substantial discretionary powers 'to have regard to the competing objectives, drawing on its ability to assess community views and needs, and to

monitor developments' (Senate 1992 p. 9). As the explanatory memorandum to the Australian Broadcasting Services Bill 1992 explains, discretionary powers are needed because: 'it is recognized that there are
tensions between the objects [of the Act] ... [and] the relative importance of an object may be determined by the issue being considered at the
time' (p. 9).

Regulation needs to apply general principles to particular issues case
by case and needs to steer an optimal path between contradictory policy
goals. It frequently requires the exercise of judgement in order to trade
off between rival interests. Under such 'fuzzy law' arrangements, the discretion enjoyed by Ofcom would help to ensure that specific political
considerations and patronage did not shape regulatory decisions. This
will require a clear identification of the regulatory goals. Parliament's
sovereignty would be maintained through its ability to amend the foundation statute – a Communications Act – which embodied the principles
and objectives of communications regulation.

But beyond that, regulation should be independent of Parliament. It is
vital that the regulatory authority be strong enough to impose and
enforce decisions that compromise the profitability of powerful firms.
But it is unlikely that a regulator will be able to do so without very firm
support from government. It is therefore important that the regulator is
independent of as well as being supported by Parliament.

At present, arguing for increased regulatory discretion means increasing power for an unelected and imperfectly accountable arm of the state.
The relation between regulator and government presupposed under a
regime of fuzzy law transfers power away from the elected and accountable. The unrepresentative and unaccountable character of UK regulation
must therefore be changed.

Self-regulation

Thus far our discussion of media and communications regulation in the
UK has focused on statutory regulation and on how sector-specific regulation and general competition law should best relate. We turn now to
the role of self-regulation: that is, the regulatory bodies established by
particular sectors of the industry. Firms jointly establish self-regulatory
bodies to achieve shared goals more effectively than they could by working individually. Self-regulatory bodies are generally independent of the
influence of any one firm but tend to reflect the collective interest of
firms in the sector they regulate. There is an inverse relationship between
the effectiveness and independence of self-regulatory bodies, as their
ability to act depends on the consent of regulated firms.

None the less, media self-regulation has notable achievements to its credit. The self-regulatory agencies' codes of conduct and complaints handling procedures provide a basis on which to build effective means to secure redress. Expert judgement is required to draw up and amend codes of conduct and to secure acceptance for them from relevant interests. Recourse to the courts is expensive and tardy and, as Baldwin observes, 'Many commentators believe prosecution is an inefficient method of enforcement compared to negotiated compliance-seeking' (1995 p. 22). Although we do not share the government's 'instinctive preference for self regulation' (Department of National Heritage 1995 p. 31) we acknowledge the value of the self-regulatory bodies' accumulated expertise. Their experience is likely to become more important as the number of firms in the media and communications sector grows. Growth makes established mechanisms of conditional licensing excessively cumbersome and likely to pose regulators with insuperable problems of compliance and monitoring. The Radio Authority's recognition of its reliance on consumer and competitor complaints to identify impermissible behaviour foreshadows inevitable developments in other media sectors. The co-operation of the statutory bodies (Independent Television Commission and Radio Authority) and the self-regulatory one (Advertising Standards Authority) is another case in point. It would be valuable for this to continue.

There are also examples of unsuccessful self-regulation. There is substantial public and Parliamentary concern about the perceived inability, or unwillingness, of the Press Complaints Commission (PCC) to check perceived abuses by the newspaper industry. Similar issues apply to the other self-regulatory bodies. For example, although most complaints to ICSTIS concern trading standards rather than content matters (ICSTIS 1995 pp. 16–17), it cannot be satisfactory that private communications are subject to regulation by a body invented and funded by the telecommunications carriers. The difficulty of making delicate judgements on content matters points to the need for greater legitimacy, transparency and accountability in all the bodies charged with such responsibilities.

The established working practices and accumulated expertise of the self-regulatory agencies should not be lightly abandoned. This expertise can usefully contribute to the formulation of the content codes on which Ofcom's decision making and sanctions will be based. Conformity to content codes is likely to be best achieved through consultation and collaboration with regulated firms rather than through an adversarial regulatory style. Ofcom should therefore consult with industrial bodies in the formulation of codes and seek their assistance in securing compliance. This should help to minimize recourse to statutory sanctions against offending firms.

Regulatory process

The Annan Report (1977 pp. 54–70) on the future of broadcasting recommended changes to regulatory procedures and institutions so that regulation was made more transparent and more responsive to public concerns. The changes proposed by Annan are generally applicable and are:

- public hearings on the performance of broadcasters and regulators every seven years
- creation of a Telecommunications Advisory Committee to advise on all electronic communications
- regular public hearings to consider the performance of broadcasters and to consult on policy initiatives run by a newly created Public Enquiry Board for Broadcasting.

Implementation of Annan's proposals would do much to ensure greater openness in regulation and policy making. It would ensure, as a former Australian competition regulator stated, 'a greater opportunity for the interests of the community to be properly aired' (Baxt 1992 p. 17).

Achieving open public processes of decision making, more effective consultation and representation of the public in these processes, and better accountability of service providers to the public whom they serve, requires procedural reform in UK communication regulation. We therefore propose that:

- the statutory UK media and communications regulator(s) be required to hold public hearings before major decisions are taken
- there be a right of representation accorded to individuals and groups with an interest in the forthcoming decision
- the regulator(s) be required to give reasons for decisions
- a Consumer Council for Media and Communications be established to advocate the consumer interest to regulators, to initiate proposals, to conduct research and to advise complainants (see Sargant 1993).

Our case for a Consumer Council for Media and Communications is strengthened by research of the National Consumer Council (1994 p. 1) which shows that consumers are not satisfied: television came bottom of all public services in terms of consumer satisfaction with choice, price and quality. Ofcom should be required to take into account the representations of this body and to comment on them in its reasoned justification for its decisions. Naomi Sargant drafted proposals for a broadcasting body, and her outline provides an excellent basis for the establishment of a body with a more general competence. The general remit for such a body would be to protect and promote the viewer and listener interest in all forms of broadcasting. Four groups of activities would be particularly important:

- responding, on behalf of a clearly defined consumer interest, to proposed policy developments, whether advanced by government, regulators or broadcasters
- initiating proposals for change, on the basis of expected future trends or problems not identified by others
- carrying out relevant research or commissioning it from others, and encouraging the publication of, and using, research by other bodies
- handling complaints from both individuals and groups, whether as consumers of programmes or as the subjects of programmes, and publishing summaries of findings (Sargant 1993 p. 166).

We endorse Baldwin's (1995 p. 11) proposal to establish a House of Commons Select Committee on the industries. The Select Committee might appoint *some* of the members of the Consumer Council. We believe our proposals for the selection and appointment of the BBC's Governors in chapter 7 provide a suitable model for selection and appointment of members of Ofcom. Proposals have consistently been made, and buried, throughout the history of UK media regulation.[12] Analogous organizations, albeit without the statutory legitimacy enjoined by UK proponents of such bodies, exist in other countries. The work of the Communication Law Centre in Australia and the Citizens Communication Center in the USA testifies to the usefulness of bodies dedicated to the identification and representation of the consumer interest.

We do not believe there is much to be gained by forcefully advocating *our* model at the expense of those proposed by others, described above. The crucial issue is commanding support for an independent, accountable and representative regulatory body. Any of the models proposed above, alone or in combination, would be preferable to the present regulatory arrangements.

Summary

1 Media and communications laws and regulation cry out for reform. Changed circumstances and technology make established institutions and instruments no longer effective. There are too many regulators,

their remits overlap and their decision making procedures are neither open nor accountable to the public.

2 Regulation is needed. We need it to ensure economic and social goals that the market may not deliver. We need it to monitor the power of private and public gatekeepers.

3 Regulation should aim at achieving four public policy goals: security, opportunity, democracy and fairness. Auditing the UK system according to these criteria shows a mixed picture of failure and success.

4 Technology allows cheaper access to information. Copyright laws restrict access to protected material for far too long. There should be new laws of 'freedom of information' extending public access to government information. New laws on privacy should protect personal privacy, qualified by the public right to know about the lives, behaviour and values of public figures in so far as that knowledge is relevant to their public office. A right to information should co-exist with a right to privacy.

5 Competition policy in the UK needs strengthening, to counter effectively the power of concentrated and strong media firms. The spirit of EU anti-trust provisions should be translated into national law to bring the UK in line with its European partners. But competition policy cannot secure goals of universal access to democratic media and affordable telecommunications services, which are positive freedoms and require active policies to be realized.

6 Convergence and change pose new problems for regulators. They blur the distinction between public and private and make current systems of regulation unevenly enforceable across different media. We propose that the confusing and wasteful plurality of regulators in the UK be reduced to a single statutory regulator, Ofcom. Ofcom should have overall responsibility for the structure and ownership of media and communication markets to promote effective competition and guard against concentrations of power. The relationship between content and carriage under this regime would require a single regulator familiar with both. Enforcement of penalties arising from offences in the domain of content would be more effective because of the regulator's leverage in the domain of carriage.

7 Common principles in content regulation will be applied across different media in proportion to the public character of the media and the possibility of involuntary consumption. Ofcom should continue to discharge the function of a 'poor man's libel court' currently undertaken by the Broadcasting Complaints Commission but with its remit extended beyond broadcasting. There should be no prior restraint or suppression of the publication of text. Access to offens-

ive material could be appropriately restricted in ways compatible with the provisions of the European Convention on Human Rights.

8 Carriage regulation should be informed by the principle of intervention, where necessary, to make competition work, to prevent the abuse of a dominant position by firms, and to secure the achievement of social goals, within a broad remit defined by Parliament based on the goal of universal service at affordable cost.

9 Ofcom must be independent of government but able to win its support. Decisions should be taken by a commission rather than by a single individual. The new relationship between government and regulator is framed by 'fuzzy law' general principles of regulation defined by Parliament which leave ample discretionary power to Ofcom. Greater powers for the regulator must be balanced by greater accountability to the public, as well as more open processes of appointment and decision making. Regulation should be open to participation and to scrutiny. Regulators will be strengthened and the process of regulation improved by the requirement for regulators to give reasons for their decisions. They will be less vulnerable to pressure from government and other power centres. Consistency in decision making will provide a more predictable environment for business.

10 Government and regulator will together ensure that a universal service obligation in content is discharged and that in free-to-air broadcasting and in other appropriate media UK citizens have universal access to high quality information, entertainment and enlightenment. Government and regulator will seek to promote choice between channels, information sources and content through diversity in content and ownership of information media.

11 Ofcom's remit shall be informed by the social justice principles set out in the report of the IPPR Commission on Social Justice (security, opportunity, democracy, fairness). Media policy shall be informed by the principle of encouraging and ensuring that UK citizens and consumers have ready access to a choice of authoritative and impartial sources and to investigative, iconoclastic and innovative material, even if this will inevitably offend some people.

12 A Consumer Council for Media and Communications should be established.

Conclusions

The complex ensemble of changes brought about by printing with movable type has been named by Marshall McLuhan *The Gutenberg Galaxy* (1962). We might refer to analogous changes brought about by electronic reproduction and transmission of information as the 'Marconi galaxy'. Both the Gutenberg and Marconi galaxies required novel institutions to realize the potential benefits of the new technologies and to balance the economic interests of producers with the interests of users and society at large. Public libraries, publicly owned mail and telephony networks, public broadcasting, and copyright laws all captured some of the gains from the new technologies and their commercial exploitation for society at large. The key challenge has not changed: balancing the benefits conferred by private ownership of intellectual property and the means of information production, distribution and exchange with realization of citizens' entitlement to information and communication. Now, as then, novel solutions are required.

Neither the old left nor the new right has been able to balance the interests of producers and consumers of media and communications. Neither has been able to translate the benefits of innovation, efficiency and a thriving industrial sector, which have been conferred by private ownership and competition, into fuller entitlements to participate in social, economic and political life for all citizens. Accordingly, there remains an indispensable place for the regulation of media and communications and for public intervention in media and communications markets. Public sector media should provide a threshold of quality below which commercial service providers sink at their peril. But the public interest will not be served by rolling back the last two decades of liberalization.

Media and communications markets are likely to remain 'failed markets'. But, market failure can provide unique social benefits – like adding an extra consumer to broadcasting at no extra cost – which may be forgone if these sectors are forced to fit into neo-classical economic paradigms. Competition and markets have a valuable role in securing the public interest in media and communications but they cannot replace regulation. If the UK is to foster a thriving, internationally competitive, industry *and* enable all of us to enjoy its benefits, a new media and communications regime is required.

A regime based on sector-specific licensing of individual firms, detailed conditions of licence and detailed monitoring of performance is tantamount to de-regulation, if the industry has hundreds of licensees. The explosive growth in the number of firms providing services following liberalization means that sector-specific licensing cannot be retained. A second key feature of the UK regulatory regime, sector-specific regulation by an alphabet soup of regulatory agencies, is incompetent to deal with the challenges of the past – let alone the future. Instead, general principles for governing the media and communications sector should be defined by Parliament. Their detailed implementation and enforcement should be delegated to a single regulatory agency, Ofcom.

We need primary legislation by Parliament to establish a Freedom of Expression Law and to incorporate the European Convention on Human Rights into UK law. We need a tort of privacy and a prohibition competition law incorporating the competition provisions of the Treaty of Rome. Ofcom will regulate the media and communications sector in order to implement the general provisions defined by Parliament. Its mandate should be to secure universal access at affordable cost to the information and communication media necessary to full participation in economic, social and political life. Achievement of this goal presupposes an economically successful UK media and communications sector. Ofcom should establish general requirements and codes of conduct within the context of which any and all firms may do business. There will be circumstances in which Ofcom will be required to make case by case decisions and to require specific behaviour of particular categories of firms. We have identified the policy issues which require new arrangements if the public interest is to be secured and have proposed a basis for the necessary regulatory and legislative arrangements. We summarize them as follows.

Universal access

Both broadcasting and telecom services can be supplied today with greater efficiency and at considerably less expense than in the past.

Genuinely universal service at affordable cost is now possible. All can, and should, have access to the means to communicate and to the information required for full social and political participation at affordable cost. In telecommunications and broadcasting the content of this 'universal service obligation' is dynamic. Once, universal service in telecoms meant access to voice telephony alone. Now it means access to a digital network. Citizens' entitlements to information, education and entertainment will continue to provide a rationale for public broadcasting services. But the role of public service broadcasters should not be confined to this definition, nor can the entitlement to information, education and entertainment be realized through broadcasting alone.

A vigorous, efficient and well funded public broadcasting system is needed not simply to overcome market failure by providing 'merit goods' (programmes which commercial broadcasters would have no incentive to provide but which are beneficial to society as a whole). It can improve the operation of the broadcasting market as a whole by providing high quality thresholds. The important public service role of setting high quality standards could be well employed in new media markets, where standards are not yet established. On-line information is probably the best example: we need an equivalent to the BBC on the World Wide Web.

Freedom of expression

The principles of freedom of thought and expression characteristic of the European Enlightenment have become globalized. Fax, electronic mail, communication satellites have made it impossible for political authorities to deny their citizens access to external information without denying the benefits of modernity to the society as a whole. Article 10 of the European Convention on Human Rights underpins European citizens' rights to free speech. In the UK, freedom of expression is not protected by law but is restricted by many Acts of Parliament. This basic principle of democracy should be guaranteed by law in a UK Freedom of Expression Act and by incorporation into UK law of the European Convention on Human Rights.

Here, we support the provisions of the Williams Report which emphasizes restriction of access to offensive material rather than suppression or censorship. We propose a sliding scale of severity in restricting access to offensive material, depending on the degree of openness of the media carrying offensive content. What may well be acceptable in adult magazines may not be acceptable when displayed on street posters. Restriction and suppression should be based on evidence that the offensive material has caused harm: what constitutes an offence will always be a matter of judgement. Society's standards should not be those of the most easily offended.

More democratic regulation and better representation of consumer interests

Who defines the public interest in media and broadcasting? Who decides what is offensive? Who decides the profitability of firms operating in imperfect, regulated, markets? How well do the gatekeepers represent British society? *Prima facie*, regulators are unrepresentative and unaccountable. Those who decide in statutory bodies are appointed by the government; those who decide in self-regulatory bodies are appointed by the industry they regulate. A new mechanism for appointing regulators is needed to ensure that those who take decisions for the UK public are more representative and accountable.

UK media regulation is poorly adapted to representing consumer interests because the UK paternalistic approach established regulation to protect consumers from themselves. A complex and fragmented regulatory system makes it difficult for consumers to know who to complain to and how to complain. Improving consumers' access to information, representation and redress is an important goal. We propose that a Consumer Council for Media and Communications should be established independently of Ofcom to identify and advocate consumers' interests. Ofcom should be charged with regulating in the public interest and should be required to consult the Consumer Council when making decisions.

Simpler and more comprehensive regulation

The responsibilities of what Michael Redley, Secretary of the Independent Television Commission, called the 'riotous mixture of exclusive and non-exclusive regulation' (IPPR's Seminar on Law and Regulatory Issues on the Superhighway in September 1995), are neither clearly defined nor comprehensive. Who regulates a newspaper on the World Wide Web? From whom does a complainant seek redress for misleading reporting in newspapers which do not adhere to the Press Complaints Commission? In a communications world where the borders between different sectors are blurring, and where the same delivery systems are used to carry sound, image and text, the existence of separate bodies in an increasingly integrated market is artificial, outdated and counter-productive.

As the communications sector converges, so too should the regulatory bodies. A single regulator charged with both carriage and content could effectively counteract, in terms of power and resources, the giant companies it is called to regulate. Its authority over carriage will give this body, Ofcom, leverage when regulating content. Its ability to provide appropriate content rules is increasingly dependent on understanding of the characteristics of different delivery systems. Responsibility should rest unequivocally with Ofcom. Self-regulatory bodies can continue to provide useful expertise by co-operating with Ofcom in drawing up codes of conduct. Reasons must be given for regulatory decisions and the opportunity to give evidence at public hearings provided for all interested parties. Giving reasons and promulgating clear decisions and transparent rules will increase the confidence of industry and the public in the regulatory process.

Policy goals should be defined by Parliament and implemented by the independent regulator

Communications and information pervade the economic, political and social life of the country. They are a key infrastructure, cost component and export sector for UK plc. British media have been deemed responsible for swinging moods and votes at election times. The quality and security of our lives depend on our ability to communicate effectively with our family and friends. Despite the sector's importance, vital decisions are made by unelected regulators who effectively decide the composition and relative importance of public policy goals. If civil society in the UK is to be strengthened and democracy extended, media and communications must be accountable to elected representatives of the public. In fact, the accountability of media and communications regulators has fallen into a kind of 'black hole'. Two contradictory imperatives are held to be true: media and communications are so important that government should not control them; and media and communications are so important that government cannot ignore them. There is one way to break this impasse: to appoint regulators by democratic mechanisms independent of government.

Defining policy goals for media and communications must remain the domain of Parliament. Government policy for the sector should be rooted in the principles of *security, opportunity, democracy* and *fairness* identified by IPPR's Commission on Social Justice (see introductory chapter). These principles should guide regulators whose discretion should be exercised independent of government. The public interest should be served without regard to the political power of the moment. The appointment of regulators should no longer be through Prime Ministerial patronage.

Self-maintaining industry structure via competition

Liberalization has created a large number of innovatory new firms and has revitalized incumbents. Still, media products are not easily substitutable, economies of scale and scope are significant, broadcasting is a public good. Formerly, concern focused on firms' size and market share. Size remains potentially important, as we recognize in proposing limits to the concentration of ownership in mass media sectors, but is no longer the main source of market power. Indeed, firms' control of bottleneck facilities is the main threat to effective competition in media and communications. Accordingly, policies based on prohibition of anti-competitive behaviour and provisions to guarantee third party access to essential facilities should take a central place in regulation.

However, competition policy can deliver markets that function well, and where concentration of ownership is not a worry, but it cannot guarantee the emergence of diverse voices representing various segments of society. An overall cap on media share by any one owner would move the system in the right direction, away from antiquated and rigid limits on cross-ownership and towards a forward looking, technology neutral system. Unfortunately, there are challenging problems in defining relevant markets and calibrating what are likely to be flexible yardsticks. And pluralism in ownership does not necessarily achieve diversity in media content. This goal can be effectively pursued by strengthening freedom of access to information, supporting editorial independence and fostering vigorous and iconoclastic public broadcasting.

International co-operation to secure policy goals in competition and content regulation

Until now scarcity of radio frequency spectrum has made it possible for the state to impose and enforce content regulation as a condition of licensing. Satellite broadcasting, the growth of international telecommunication networks and digitalization have mitigated spectrum scarcity and extended the supply of broadcasting services from overseas. Thus media and communication regulation faces new difficulties of jurisdictional definition and of enforcement.

The end of national control of media and communications can be seen either as marking a positive emancipation of media and communications from the tutelage of the state or as a dangerous slide towards anarchy. Undoubtedly there is bit of both. The same anarchy that allows Dutch liberals to teach the rest of the world how to grow marijuana, or emancipated Scandinavians to beam porn to the UK, also permits persecuted Kurds to communicate with their people from outside the Kurdish region or enables critical information to circulate in authoritarian Gulf states. But, despite enormous differences in matters of contents and ethics between sovereign European countries, co-operation between EU member states can provide individuals and organizations with more effective redress for abuse than they currently enjoy. Closer collaboration with EU partners is necessary, rather than just desirable, for successful regulation of the modern communications sector.

Economically successful UK industry

A world of increasing dependence on communications and increasing trade in information offers the UK unique opportunities. Works in English circulate more easily than works in other languages and the English language market is bigger and wealthier than other language markets. Even if this does not imply that English language producers will always succeed, there are opportunities for English products and producers which do not exist for other language communities. To make the most of these opportunities UK industries must be able to adapt to conditions which are changing very rapidly and assume an economically optimal shape – subject to satisfying other public policy criteria. For example, if it is desirable to have regional news available on radio and television in all parts of the UK, or indeed to have production of television or radio programmes in particular locations, then the firm which is best placed to provide these services – to the regulator's satisfaction – should be permitted to provide a broadcasting service in the area in question, whether or not it is a separate regionally based company.

In spite of the erosion of national communication sovereignty through technological change, the UK government still has significant powers to shape the structure and performance of media and communications in the nation – as we have argued in our discussion of rival scenarios for managing the transition to digital broadcasting (chapter 2). But it should not be forgotten that the government also disposes of considerable power as one of the largest and most influential customers for communication services and as an important source of information. Its market power as a customer should be used to foster best practice, open standards, interoperability of systems and the like. Underpinned by freedom of information legislation, government tenders should be characterized by transparent rules and criteria for awarding contracts. Purchasing from 'pro-competitive' firms is one of many informal ways to influence outcomes open to government to influence outcomes. A dynamic and competitive market nurtures national champions far more effectively than cocooning home-grown giants. Moreover, making government information widely available in the new electronic on-line libraries is likely to effectively stimulate the development of the UK as an information society.

Appendix 1
The Telecommunications Sector: Economic and Technological Background

Technological characteristics of the telecommunications network

Telecommunications consist of the carrying of voice, data or image signals between different customers along a network. There are two large components to the sector: the manufacture and building of a telecommunications network and the provision of services over it. The same operator can be responsible for both laying the network and providing the services (network operator, for example BT) but provision of services over other operators' networks (service provider, for example AT&T in the UK) is increasingly common.[1] The two components of the telecommunications sector are strongly interrelated, in the sense that the technical features of the network determine the assortment of services provided and their cost while the volume and character of services carried over the network determine its economic viability. A telecommunications network is composed of several distinct components:

- the telephone sets (the end-user equipment)
- the cable or other link connecting them (transmission equipment)
- the exchange, an intelligent device which allows calls to be routed to their destination without each caller being directly connected to every other one (switching technology).

Figures 6 and 7 illustrate the importance of the switching technology which lies at the core of the telecommunications technology. In figure 6, three customers in country A are connected to each other and to three customers in country B using telephone lines without exchanges. Each house needs five cables, three of which are laid underwater to connect customers between different countries. In figure 7 the efficient way to create a network is to group all customers in one given local area and connect them to one local exchange. The wiring that connect households to their local exchange is called the local loop. Each local exchange is then linked to a main exchange and on to a trunk line. In practice, since efficiency requires not only the shortest route, but also reliability, the hierarchy of switches is slightly more complex, and any two customers located in different local exchanges can be connected through at least four different routes, in case of heavy traffic or faults in any individual exchange box. Different areas of the network are characterized by different economic features, depending on how costly it is to lay them and how much traffic goes through them to pay back the investment. For example, the local loop is still considered a natural monopoly by some, although this belief is increasingly challenged. Whether it makes sense to

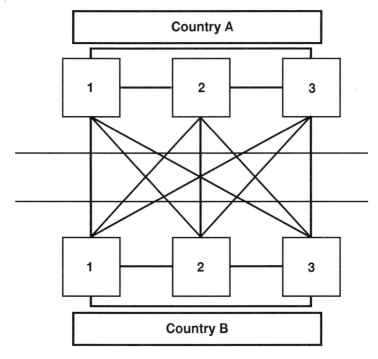

Figure 6

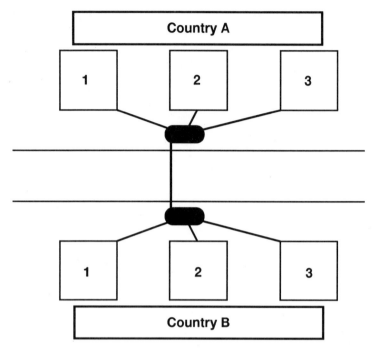

Figure 7

allow costly duplication of infrastructure at the local level depends on the traffic volume expected to exist in that segment of the network. Competition in the local loop is already encouraged in areas characterized by very high traffic, like the City of London, or where economies of scope are significant, like the cable television industry. Competition in trunk lines, where all long distance and international traffic is routed, is more established: transmission and investment costs keep falling while traffic increases.

The main technological developments occurred in the switching and transmission technology. In the 1970s a new transmission technology, optical fibre, was substituted for copper wires and offered huge savings, as it has far higher capacity and is less cumbersome to lay. It is estimated that the cost of laying submarine cable for intercontinental communications has decreased by a factor of 150 with the use of the optical fibre technology. At local level, optical fibre is used up to the local exchange, but usually not past it, as normal copper wire has sufficient capacity to carry basic voice telephony. However, optical fibre is needed for more intensive use of the network, as in transmission of images and data. Much of the current discussion about the economics of network building centres around the question of whether it is economical to substitute copper with fibre all the way up into the access network, i.e. to the customer's home (fibre to the home, FTTH). This is currently not the case even when television and telephony applications are used jointly (Department of Trade and Industry 1994). A viable alternative, also used by the cable companies, is to run fibre to the kerb (FTTK) – or to the building (FTTB) where there is multiple occupancy – and copper to the individual sockets.[2] Transmission of telephone signals can also be obtained without wires. Satellites provide transmission links across continents. Radio links can be used both for fixed telephony and for mobile communications. Fixed radio links transmit to a fixed antenna while mobile cellular technology can transmit directly to the handset.

The development of digital switching technology started the real revolution in telecommunications. Digitalization implies better quality and greater reliability than the traditional electromagnetic exchanges at a lower price.[3] Crucially, it means that voice signals can be treated in the same way as computer data, thus opening the way for the convergence of IT and telephony. This explains how the basic voice telephony service could be complemented by teledata services (fax and modem services) and will eventually include images (video phones, video-on-demand). Digital technology in the switches has greatly increased the capabilities of the exchanges: the origin and destination of calls can be recorded and the most efficient route can be automatically selected.

The choice of technology to be used in the network for low volume users (residential and small businesses) depends crucially on its cost. The cost of transmission technology has been falling faster than that of digital switches and hence the network is developing with fewer switches and longer lines. BT claims (see Armstrong, Cowan and Vickers 1994) that switching costs are increasingly access related, i.e. dependent on the number of lines connected rather than the volume of calls passed through. This implies that the cost of telephone services has a large fixed component, and is not so sensitive to usage, as our bills imply, or to the necessity to accommodate high capacity for peak time, as others argue. Identifying the costs of telecommunications services is a key task for the regulator, and attributing cost variations to one factor or another has very important implications for the way new operators are allowed to use the existing network. To complicate matters even further, the cost structure of a network is still capable of changing dramatically with technical change.

Economies of scope with existing networks of other utilities are a relatively new idea. Privatized electricity companies have sought to diversify into telecommunications since the beginning of the 1990s, by establishing joint ventures with telecom operators. They are able to build networks cheaply by wrapping telecommunications cables around their electrical wire networks. The largest project of this sort, Energis, a telecom venture of the National Grid, entered the market in 1994. Other significant projects include Ionica – a joint venture between, among others, Kingston Communications, Yorkshire Electricity and Northern Electric – and Torch, formed by Yorkshire Electricity and Kingston Communications. Another example is Tele2, Sweden's second largest telecom operator, which uses a network based on the fibre capacity of the national railway company and other utilities. (Public Network Europe 1995).

Economic characteristics of the industry: barriers to entry

Why do we worry about barriers to entry? In theory, any market where producers are able to make profits attracts new entrants. As new competitors undercut the incumbent's price, profit margins for the entire industry go down until they reach zero. In practice, however, entry in profitable markets can be difficult because of the characteristics of the production function (for example, heavy investments in research, or significant economies of scale) and because of the ability – given those characteristics – of the firm already engaged in production, the incumbent, to deter entry. Both types of barrier apply and often interact in the UK telecommunications market. For example BT's network constitutes an enormous absolute advantage (huge investment already sunk) which the regulator sought to eliminate by requiring the incumbent to guarantee access to new entrants. BT, however, can still use this absolute advantage as a strategic tool to deter entry, by raising the price or reducing the quality of access to its network. Relevant characteristics of the telecom sector that can constitute a barrier to entry are the following:

High R&D costs   Telecommunications is a highly technical and dynamic sector, where huge sums are spent in developing the products and the technology needed for producing them. However, service providers can enter the sector with minimal R&D expenditure, using intelligent networks built by others.

Sunk costs   Networks require huge investments. If the firm which built the network was to exit the market, that investment could not be put to any other use. It hence represents a sunk cost which cannot be recovered, and commits firms to stay in the market. Sunk costs that are not recovered are a net loss to society: nobody can use the costly infrastructure. Hutchinson's Rabbit, a failed attempt to build a telepoint network in the UK, is a case in point: hundreds of millions have been written off. In a market with no sunk cost, entry can happen fast and prevent the incumbent from reacting quickly by varying price ('hit and run' competition). The existence of sunk costs is hence an important barrier to entry.

Economies of scale Economies of scale exist when the cost of producing a unit of a given product decreases with the amount of units produced. In any market characterized by economies of scale, a potential entrant faces higher costs for producing the same good until it reaches a certain scale of production (the minimum efficient scale) which allows it to undercut the incumbent's price. This implies two types of disadvantage for the entrant: the difficulty of raising enough capital to finance a large scale product and/or the penalty of producing at below efficient scale (Geroski 1991).

Economies of scope The telecommunications services market is a multi-product market: voice telephony, data transmission, leased lines, private circuits. Production in such a market is often characterized by economies of scope, i.e. the possibility of using the same input for more than one output, so that producing a range of products together is cheaper than producing them separately. Economies of scope can also offer advantages for the new entrant, as is the case for operators which are able to use the networks of other utilities for telecommunications.

Network externalities The value of a network is higher the more people are attached to it. Every time a new customer joins in, the value of the network increases for all the customers already connected. Without interconnection agreements small networks are very strongly disadvantaged.

Natural monopoly elements The economies of scale discussed above can apply to the whole industry as well as a single firm. In other words, the cost of the entire industry can be lower if production is concentrated in just one firm. Elements of natural monopoly exist in international telephony, until investment in transmission is recovered on international networks. Provision of local telephony is also considered a natural monopoly. This view is now controversial, since competing technologies and economies of scope have been successfully used in the local loop. It is now possible to argue that the inefficiency of duplicating local lines is offset by the gains in lower prices and greater innovation that competition is likely to bring.

By themselves, the characteristics of the telecommunications industry we have just discussed are likely to restrict entry, rather than prevent it. It is the interaction between the incumbent firm's behaviour and the market characteristics which creates real barriers to entry. In oligopolistic markets like telecommunications, the firm which first entered the market is in a position to pre-empt and discourage even a more efficient and innovative new entrant. The incumbent can use its cost advantages, for example scale advantages, to engage in predatory pricing. Example of first mover advantages are the following.

Switching costs   A new firm entering the telecommunications market has to convince customers to abandon their previous provider for the new one. In the best case, where the new operator is perfectly compatible with the incumbent and has access to its network, the entrant has to win against customers' inertia. A recent Oftel survey[4] (*New Media Markets* 1995) proved that customers' inertia is the main obstacle to cable telephony penetration.[5] Only those customers who are dissatisfied with the incumbents' product will initially want to go to the trouble of trying the new one. Many others who would benefit from the new services on offer may not be willing to try. Given that telecommunications services are 'experience' products, i.e. products whose quality cannot be judged before purchase (or without acquiring considerable information), winning customers' inertia can impose high costs on new entrants. Customers who want to switch operators often face some actual cost, such as having to change their telephone numbers and having to buy new, more expensive equipment. In the UK, customers who wanted to switch to Mercury for their long distance calls had to buy expensive 'blue button' phones or 'smart sockets', where the complex access code to Mercury's trunk lines was memorized. Eventually, the access code was simplified to a few digits to be dialled before the phone number, but this still constitutes a confusing hurdle to the usage of the rival's network. In the Hull area, where Kingston[6] revised its numbering system to impose a pre-dialling code for both Mercury and BT, Mercury's penetration rate was much higher than in the rest of the country.[7] At the time of writing (January 1996) number portability has just been introduced after twelve years of competition, battles between Oftel and operators and a referral to the Monopoly and Mergers Commission.

Advertising   Advertising expenditure is crucial for overcoming customers' inertia. It can be a barrier to entry in many different ways. The new entrant needs to advertise and inform in order to gain any market share, particularly when selling experience goods. According to the Oftel survey (*New Media Markets* 1995) three out of ten businesses in areas served by cable telephony were unaware that cable telephony was available.[8] Unlike the incumbent, the rival is facing consumers who already have an alternative they are familiar with. The threat of entry may induce the incumbent to increase its own advertising, which can raise the price of advertising overall and for the entrant. BT began to advertise its public phones only after announcements of a new entrant in that segment, with a 'pop video doubling as a BT ad' (*Guardian* 16 October 1995). A firm already established in the market can enjoy scale advantages in advertising expenditure:

- Advertising rates decrease with bigger overall budget.
- There are 'minimum efficient' thresholds for advertising to be effective.
- The effectiveness of different media increases more than proportionally to the increase in price.

Vertical integration   Regulation of the telecommunications sector does not always result in the dismantling of a vertically integrated monopolist. Keeping a vertically integrated incumbent imposes the burden of devising efficient access charges. On the other hand, separation can be inefficient if there are economies of scope within different activities that cannot be successfully arranged via long term contracts. In the UK, the decision to keep BT's structure intact was dictated by political pressures for a quick privatization rather than economic considerations. A vertically integrated firm has control over some of the inputs to the rival's end products, so the regulator has to make sure that those inputs are offered to the entrant at the same conditions available to the incumbent firm. Provided that interconnection charges are successfully regulated, the incumbent can still operate non-price discrimination against the rival's customers. It could, for example, route the rival's traffic into the most obsolete and faulty segments of its network, increasing the amount of call failures and traffic related delays. Delaying an agreement on interconnection charges might be another way to use the incumbent's power to the detriment of the new entrants.

Access to retail and support infrastructure   If the incumbent firm has control over the infrastructure supporting production and sales, it can enjoy considerable cost advantages over its rivals. BT, for example, retained a great deal of control over strategic assets that support both sales and production. It holds critical information on the entire customer base through its billing and the telephone directory, to which it has exclusive rights. Pay and public phones offer another example of first mover advantage in the retail/support side: all the best locations for a public phone are already equipped with a BT phone, and Mercury can only rarely offer a public phone service which is not in direct proximity to that of its rival. Mercury adopted the strategy of differentiating the product it offered alongside BT's phone boxes (i.e. credit card public phone services), but BT could imitate at lower cost by upgrading its existing public phones. Geroski (1991) observes that these types of advantages often have similar effects to scale economies, in that they raise rivals' fixed costs.

Capital requirements.   The ability of a new entrant to raise the necessary capital depends greatly on the risk it faces in attempting entry. The incumbent has an incentive to engage in predatory behaviour in order to increase the perceived risk of entering the market. This will have the effect of raising the cost of capital for the potential rival.

However, being a first mover is not always and invariably advantageous, and this is particularly true of industries that are characterized by fast technological change. For example, a newcomer can today leapfrog technology with much lower costs than, say, twenty years ago. The incumbent has to go through a costly process of substitution and change of physical assets as well as production and management systems.

Appendix 2
The Gatekeepers: Who Regulates Media Content?

In 1995, IPPR researched the composition of the regulatory bodies in charge of the content of media and telecommunications services and products. The research was based on publicly available information, annual reports and telephone interviews. Its findings are bound to be only indicative at the time of publication, as new regulators are appointed and old ones retire. The main results are summarized in table 9 (given in chapter 8, p. 160). This appendix gives greater detail on the institutions. We gratefully acknowledge the co-operation of all institutions surveyed and Adam Jacobs's research on which this appendix is based.

Overview

In 1995 there were a total of 90 'gatekeepers', i.e. members of the decision making boards of UK regulators. A slight majority are male, 53 gatekeepers (or 57 per cent of the total), and only 4 members are not white (4 per cent). Although it is extremely difficult to provide accurate figures, all the committee memberships are London dominated, that is to say about 80 per cent of members live in London. A quarter of the total membership (22) were educated at Oxford or Cambridge University. Table 10 shows the professions represented in these regulatory institutions.

Table 10

Profession	No. of gatekeepers	% of total
Media/journalists	20	23
Business/marketing	11	13
Academics	8	9
Law	7	8
Civil servants	6	7
Teaching/education	5	6
Former politicians	5	6
Trade union	3	3
Church	3	3
Aristocrats	3	3
Consultants (telecoms)	3	3
Other	14	16
Total	88	100

Chairpersons

- Of the 11 bodies, 8 are chaired by men and 2 (ICSTIS and BSC) by women.
- Half of the bodies (5) have chairs who were Oxbridge educated.
- All chairpersons are white.
- Average age of chair (excluding S4C) is 63.
- Baroness Dean (ICSTIS) is the only chairperson younger than 60.
- The Chairpersons of ICSTIS and ASA are responsible for choosing their own members.

Committees appointed by the Government

Government appointed institutions are:

- Broadcasting Standards Council
- Oftel
- Broadcasting Complaints Commission
- Radio Authority
- Welsh Fourth Channel Authority
- BBC Governors
- Independent Television Commission.

These bodies have 47 members (excluding employees, who are not directly involved in the decision making process). The male majority is higher than the average in government appointed bodies, with approximately 60 per cent men and 40 per cent women. There are 2 non-white members (4 per cent) in total. The percentage of members who are Oxbridge educated is also higher than average in this group: 15 members, or 33 per cent of the total.

Under the Broadcasting Act 1990, the provision of payments to Commission members is at the discretion of the Secretary of State of the Department of National Heritage. The Secretary of State also determines the emoluments of the BBC Governors. The total remuneration (not including expenses) received by members of broadcasting regulatory bodies established by statute and Royal Charter in 1993–4 (or calendar year 1993) was £774,970. It is difficult to pinpoint exactly how much each authority member receives and to determine the exact number of days worked (generally these committees meet monthly), but here are some examples from the 1993–4 financial year:

- Broadcasting Complaints Commission (5 Members): Commissioners' salaries £88,048; Chair £31,450.
- Radio Authority (7 Members, monthly meetings): Members' salaries £110,000; Chair £43,950.
- Welsh Fourth Channel Authority (7 Members, 11 meetings): Members' salaries £85,000; Chair £35,000.
- Broadcasting Standards Council (8 Members): Members' salaries c.£143,000; Chair £38,303.
- British Broadcasting Corporation (15 Governors): Governors' emoluments £206,927; Chair £73,312. The Chairman's emoluments increased by 12 per cent in 1993–4 compared to the previous year.
- Independent Television Commission (10 Members): Members' emoluments (1993) £141,995; Chair £57,360.

The regulators

The following sections examine each regulatory institution in greater detail.

Independent Television Commission

The Independent Television Commission (ITC) derives its powers from the Broadcasting Act 1990 (table 11). The Secretary of State for National Heritage appoints the Chairman, the Deputy Chairman and the Members of the Commission (usually eight). They are supported by a permanent staff which includes specialists in licensing, finance, engineering, public affairs and the regulation of programmes, advertising, cable and satellite. Its remit is to:

- License commercially funded television services in the UK, whether delivered terrestrially or by cable, local delivery or satellite; license services provided from the UK to viewers in other countries.
- Regulate these services through its published licenses, codes and guidelines. It has a range of penalties for failure to comply with them – warnings, fines, revoking licence.
- Ensure that a wide range of television services is available throughout the UK and that, taken as a whole, the services are of a high quality and appeal to a variety of tastes and interests. By its nature the ITC defines 'quality'.
- Ensure fair and effective competition in the provision of these services.

Table 11   Independent Television Commission profile

Members:	10	Male 6 Female 4
Average age:	56	Chair: Sir George Russell (60). Oldest member: Pauline Mathias (66). Youngest member: Earl of Dalkeith (41)
Oxbridge education:	2	Earl of Dalkeith, John Ranelagh
Occupations:		Business/marketing 5, media 2, academic 1, aristocrat 1, headmistress 1

Miscellaneous:

Pauline Mathias: convent educated; 'against co-education and the pill for 16 year olds' (*Sunday Telegraph*)

Jocelyn Stevens (Deputy Chair): Managing Director Express Newspapers (1977–81; dubbed 'Piranha Teeth' (*Private Eye*)

Sir George Russell: non-executive Chairman of 3i Group

Jude Goffe (Afro-Caribbean): (1984–91) held several investment positions at 3i

Earl of Dalkeith: heir to two dukedoms, 258,000 acres of estates, stately homes; will inherit one of the few Leonardo da Vinci's in private hands; on the Board of Border TV

Broadcasting Standards Council

The Broadcasting Standards Council (BSC) is a statutory organization under the Broadcasting Act 1990 (table 12). It is an advisory body. It deals specifically with matters of violence, sex and taste and decency, such as bad language, coverage of disasters and issues of race and gender. The BSC is funded by grant-in-aid from the Department of National Heritage, and its eight Members are appointed by the Secretary of State, under the public appointment system. Its two main functions are considering complaints and commissioning research into public attitudes. It also has a reporting role, which it completes by monitoring developments in broadcasting and reporting trends and changes. The BSC covers all television and radio programmes and advertisements, including cable and satellite, but has no powers to preview material to be broadcast, or to prevent the transmission of programmes. Findings are published in a monthly bulletin and the broadcaster may be required to publish a summary on air or in the press.

Table 12   Broadcasting Standards Council profile

Members:	8	Male 4 Female 4
Average age:	58	Chair: Lady Howe (63). Oldest member: The Very Rev. John Lang (67). Youngest members: Matthew Parris and Sally O'Sullivan (45)
Oxbridge education:	1	
Occupations:		Media 3, teacher 1, aristocrat 1, psychiatrist 1, Church 1, politician 1

Miscellaneous:
  Church connections: Robert Kernohan, Editor *Church of Scotland Magazine*; John Lang (age 67), Dean of Litchfield (1980–93) and Chaplain to the Queen (1976–80); Dr Jean Curtis Raleigh (age 61) convent educated
  Tory connections: Lady Howe, Conservative Women's National Advice Committee; Matthew Parris, ex Tory MP; Robert Kernohan, Conservative Party Scotland

Broadcasting Complaints Commission

The Broadcasting Complaints Commission (BCC) is an independent statutory body which derives its power from the Broadcasting Act 1990 (table 13). The BCC is known as the 'poor man's libel court'. It deals with complaints about unjust or unfair treatment in a programme or unwarranted infringement of privacy. Individuals or organizations can complain, but only if they are the 'person affected' by the programme, or their representative. The BCC covers all television and radio programmes, teletext and advertisements, including cable and satellite. Complaints are referred to the broadcaster, which has to provide a transcript of the programme and a written statement in answer to the complaint. The complainant can then respond and the broadcaster can make a further submission. The Commission can also call private hearings. The Commission sends a written adjudication to the complainant and broadcaster and usually requires a summary to be broadcast and published in the *Radio Times* or *TV Times*. The summary of the Commission's adjudication is usually broadcast on the same channel and at a similar time to the programme which was the subject of the complaint. This is the only sanction available to the Commission: it cannot require the broadcasters to apologize to the complainant, to broadcast a correction, or to provide financial compensation. All appointments are made by the Department of National Heritage.

Table 13 Broadcasting Complaints Commission

Members:	5	Male 3 Female 2	
Average age:	60	Chair: Canon Peter Pilkington (61). Oldest member: Tony Christopher (70)	
Oxbridge education:	1	Canon Peter Pilkington	
Occupations:		Diplomat 1, trade unionist 1, business/marketing 1, media 1, Church/education 1	

Miscellaneous:
Canon Pilkington: House Master Eton (1962–75); High Master St Pauls (1986–92)
Tony Christopher CBE: Trustee of IPPR
Donald Allen CMG (age 64): Foreign Office since 1948
Jane Leighton: Executive Director Mersey TV (1988–91)

Radio Authority

The Radio Authority (RA) can apply sanctions to licensees who break the rules (table 14). Sanctions include broadcast apologies and/or corrections, fines and the shortening or revocation of licences. The Secretary of State for National Heritage appoints the Chairman, Deputy Chairman and five Members of the Commission. The RA derives its powers from the Broadcasting Act 1990. It is empowered to:

• Plan frequencies, appoint licensees with a view to widening listener choice, and regulate programming and advertising.
• License and regulate all independent radio services. These comprise national, local, cable, national FM subcarrier, satellite and restricted services. The last include all short term, freely radiating services (for example, 'special event' radio) and highly localized permanent services such as hospital and student radio.
• Monitor the obligations on its licensees required by the Broadcasting Act.

Table 14   Radio Authority profile

Members:	7	Male 4 Female 3
Average age:	50	Chair: Sir Peter Gibbings (66), also oldest member. Youngest member: Margaret Corringham (40)
Oxbridge education:	3	
Occupations:		Law 3, media/journalism 2, business 1, civil servant 1

Miscellaneous:
   No member has any experience of radio
   One member of Afro-Caribbean origin: Jennifer Francis
   Sir Peter Gibbings: former Chair of the *Guardian* and *Manchester Evening News* (1973–88), Anglia Television Group (1988–94); Director of *The Economist* (1987–); served in 9th Queen's Royal Lancers
   Andrew Reid: solicitor, various business interests which include horse breeding and racing; owns hotel and bar in London and has varied farming interests

BBC Governors

The BBC Governors are the focus of legal authority for the BBC, with day-to-day responsibility for securing and maintaining its independence (table 15). They are the trustees of the BBC. Governors retain final authority in all areas. They are appointed by the Department of National Heritage. The Governors appoint the Director-General and key executives, ensure that the BBC has a remuneration policy that enables it to attract executives to key positions (via their Remuneration Committee) and are accountable for the proper use of public money. The Board of Governors receives its powers to:

• Stay closely in touch with public opinion
• Ensure that the BBC's overall strategy reflects the public's needs and interests
• Monitor and review performance against agreed objectives
• Ensure compliance with statutory requirements and BBC guidelines
• Guarantee regular reporting to the licence payer and to Parliament.

Table 15   BBC Governors profile

Members:	9	Male 7 Female 2
		(+3: three vacancies at the time of research)
Average age:	59	Chair: Marmaduke Hussey (71), also oldest member
Oxbridge education:	5	
Occupations:		Civil servant 2, trade union 1, politician 1, business/marketing 2, headmistress 1, Church 1, media 1

Miscellaneous:
    All governors are white
    Marmaduke Hussey: served with Grenadier Guards; Chief Executive of Times Newspapers Ltd (1971–80); his wife, Lady Susan Hussey, Lady-in-Waiting to the Queen since 1960
    Rev. Norman Drummond (Cambridge): Governor for Scotland, Chaplain to the Queen in Scotland, former Chaplain to the Parachute Regiment
    Gwyn Jones: Welsh Governor and member of S4C Authority
    Lord Cocks: ex Labour MP Bristol South; Deputy Speaker in House of Lords since 1990
    Lord Nicholas Gordon Lennox KCMG KCVO: Foreign Office since 1954; retired from HM Diplomatic Service as Ambassador to Spain 1989; Hon. Colonel 4th Battalion Royal Green Jackets TA

Welsh Fourth Channel Authority

The Welsh Fourth Channel Authority is the regulatory body of S4C, the broadcasting service transmitted on the Fourth Channel in Wales (table 16). S4C broadcasts an average of 32 hours per week in Welsh, mainly during peak hours, and the remainder of the time rebroadcasts Channel 4's English language programmes. The Authority reports to the Department of National Heritage. The Secretary of State for National Heritage appoints the Chairman and six other Members.

Table 16   Welsh Fourth Channel Authority profile

Members:	7	Male 4 Female 3
Average age:	n/a	Chair: Ifan Prys Edwards (71)
Oxbridge education:	1	
Occupations:		Academia 3, civil servant 1, architect 1, business/banking 2

Miscellaneous:
  Geralt Jones (Welsh Film Council) and Elan Stephens (Lecturer Film and TV Studies) are the only members with media backgrounds

British Board of Film Classification

The British Board of Film Classification (BBFC) is an independent, non-governmental body set up to bring a degree of uniformity to the standards of film censorship imposed by the many disparate local authorities (table 17). The bulk of the work is done by a team of ten Examiners, which is ratified by senior management. The President and Vice-Presidents have ultimate responsibility for all decisions taken by the Board, whether on film or video. Under the terms of the Video Recordings Act 1984 they are designated as the persons responsible for all matters relating to the classification of video works. Although day-to-day decisions are normally delegated to the Director and the Principal Examiner, the Presidents do participate in this process by viewing a film or video deemed likely to cause comment, give offence, or bring the Board's reputation into disrepute, or one that has produced considerable debate within the Board itself. The Presidents act as a court of appeal for distributors dissatisfied by a given category. As far as cinema categories are concerned, they have the last word.

Table 17 British Board of Film Classification profile

Members:	5	Male 3 Female 2
Average age:	65	President: The Right Hon. The Earl of Harewood KBE (72), also the oldest member
Oxbridge education:	4	
Occupations:		n/a

Miscellaneous:
  The gatekeepers with the last word in film matters are all white, near 70 years of age and Oxbridge educated
  Vice-Presidents: Lord Birkett (66), Monica Sims OBE (*c.*70)

Independent Committee for the Supervision of Standards of Telephone Information Services

The Independent Committee for the supervision of Standards of Telephone Information Services (ICSTIS) is the supervisory body for the premium rate telecommunications service industry (table 18). The Committee is responsible for the production of the Code of Practice which covers the provision of premium rate services by means of a public telecommunications network. The role of ICSTIS is to supervise both the content and the promotional material for premium rate services and to enforce the Code of Practice with the support of the network operators. The Committee consists of around ten members, all appointed in their individual capacities, and is supported by a full time secretariat. All the appointments are made by the Committee Chair in conjunction with the network operators. ICSTIS can use a variety of sanctions, which it will apply according to the seriousness with which it regards the breach. They range from warnings to fines and prohibiting the service provider from providing premium rate services for a defined period. Its main tasks are:

- setting and maintaining standards for the content and the promotion of premium rate services, and keeping those standards under review
- monitoring those services to ensure that the content and promotional material comply with these standards
- investigating and adjudicating upon complaints relating to the content and promotion of the premium rate services, which may include the imposition of sanctions.

Table 18   Independent Committee for the Supervision of Standards of Telephone Information Services

Members:	10	Male 4	Female 6
Average age:	est. 46 Chair: Baroness Brenda Dean (52). Oldest member: Jeremy Mitchell (66)		
Oxbridge education:	1		
Occupations:	Law 2, social work 1, teaching 2, trade unionist 1, specialist consultants 3, journalist 1		

Miscellaneous:
  One member of Indian origin: Shirley Daniel
  Baroness Dean: Member of Press Complaints Commission; until 1992 was Deputy General Secretary of the Graphical, Paper and Media Union
  Jeremy Mitchell: Member of IPPR Media and Communication Programme Advisory Committee

Press Complaints Commission

Following the Report of the Calcutt Committee (1990), the Press Complaints Commission (PCC) was set up in place of the Press Council as a non-statutory self-regulatory body (table 19). A Press Standards Board of Finance (Pressbof) was put into place and charged with raising a levy upon the newspaper and periodical industries to finance the Commission. A committee of national and regional editors produced a new sixteen clause Code of Practice for the Commission to uphold. Publishers and editors committed themselves publicly to their own Code of Practice and to ensuring funding for the disciplinary body to uphold the Code. The success (or failure) of the PCC rests upon the Code which is reviewed periodically by a special committee of editors. The Commission takes the final responsibility for the Code of Practice. It is a fundamental principle of the Commission's procedure that members themselves exercise final control over complaints. In the case of complaints, the Commission aims to reconcile complainants and editors and thus to avoid the need for formal adjudication. Cases which cannot be resolved proceed to monthly meetings of the Commission for adjudication. But the Commission has no formal powers other than to publicize breaches of the Code: such is the nature of self-regulation. There are sixteen Members, currently chaired by Lord Wakeham. There are three classes of Member: the Chairman, Public Members and Press Members.

The Chairman is appointed by Pressbof (currently chaired by Harry Roche) and is to have no newspaper, periodical or magazine interests. There are eight Public Members (non-press) appointed by an Appointments Commission which consists of the Chairman of the Commission, the Chairman of Pressbof and three other independent persons nominated by the Chairman of the PCC. The Commission presently comprises the independent Chairman, eight other non-press Public Members and five senior editors from across the industry (currently two vacancies for Press Members).

Table 19   Press Complaints Commission profile

Members:	13	Male 10 Female 3
Average age:	60	Chair: Lord Wakeham PC, FCA, JP (71), also the oldest member.
Oxbridge education:	1	
Occupations:		Editors 5, trade union 1, academic 1, politician 2, civil servant 1, law 1, art 1, aristocrat 1

Miscellaneous:
   Lady Elizabeth Cavendish, LVO, JP: extra Lady-in-Waiting to Princess Margaret since 1988
   The Lord Tordoff: age 66; Liberal Democrat Chief Whip in House of Lords (1988–94)
   Professor Robert Pinker: age 64; Privacy Commissioner; Council Member ASA

Advertising Standards Authority

The Advertising Standards Authority (ASA) promotes and enforces standards in all non-broadcast advertisements in the UK and is independent of both the government and the advertising industry (table 20). The standards are laid down in the British Codes of Advertising and Sales Promotion. The codes require that advertisements should be: legal, decent, honest and truthful; prepared with a sense of responsibility to consumers and to society; and in line with the principles of fair competition. The ASA has a Council of thirteen Members, the majority of whom have no connection with the advertising industry. Its Chairman, Lord Rodgers (former Chair Sir Timothy Raison – Eton, Oxford), is also an independent appointment. The Codes are drawn up by the

Committee of Advertising Practice (CAP), which comprises the trade and professional organizations of the advertising industry, including those representing the media, advertisers and agencies. The following sanctions are possible:

- If an advertisement breaks the Codes, advertisers are asked to amend or withdraw it. If they do not, the media may invoke their terms and conditions of business which requires adherence to the Codes, and may ultimately refuse further space to advertise.
- Both advertiser and agency may also jeopardize their membership of trade or professional organizations if they persistently break the Code's rules.
- If these measures fail to prevent a misleading advertisement from appearing, and it continues to be published, the Authority can ask the OFT to invoke the Control of Misleading Advertisements Regulations 1988: these empower the Director General of Fair Trading to apply for an injunction against advertisers, agencies and media to prevent further publication.

Table 20   Advertising Standards Authority profile

Members:	13	Male 7 Female 6
Average age:	n/a	Chair: Lord Rodgers (66). Oldest member: n/a
Oxbridge education:	1	
Occupations:		Advertising 5, academics 3, social work 1, civil servant 1, politics 1, law 1

Miscellaneous:
    One member of Pakistani origin: Sheila Iffat Hewitt
    Patricia Mackesy: formerly Chair of Oxford Magistrates Bench and of Camden Juvenile Court
    Professor Robert Pinker: Privacy Commissioner for the PCC
    Professor Lord Rodgers: Labour MP (1962–81); Vice-President SDP (1982–7); Garrick Club
    Richard Bradley: Vice-Chairman of L'Oreal (UK)
    Dr Ian Markham: Lecturer in theology, Essex University; published on Christian ethics and culture
    Christopher Hawes: Chairman TSMS Group Ltd (1989–94)
    Elizabeth Filkin: Adjudicator for Inland Revenue and Customs and Excise, Director of Britannia Building Society
    Catherine Peckham: Professor of Paediatric Epidemiology

Notes

Introduction

1 John Reith, later Lord Reith, was the first Director-General of the BBC. He held office, first as General Manager then as Managing Director and as Director-General, from 1922 until 1938. His vision and values permanently marked the BBC and British broadcasting. His complex legacy left the BBC committed to 'bringing the best of everything into the greatest number of homes' (cited in Smith 1974 p. 44) but customarily defining 'the best' in somewhat narrow, patrician and improving terms. Famously, new comedians in the early days of the Corporation were given written instructions which specified: 'No gags on Scotsmen, Welshmen, Clergymen, Drink, or Medical matters. Do not sneeze at the microphone' (cited in Briggs 1961 p. 289).

2 The Party's *Arts and Media* publication of 1991 is silent on such matters. However, it made laudable, albeit cautious, commitments to public service broadcasting, to liberalization of access to information and of media content regulation, and to tackling cross-media ownership.

3 The original text, published in 1981, began: 'British broadcasting was started as a public service, unified in structure and aim. Broadcasting – monopoly or duopoly – was always justified in these terms and an assumption of commitment to an undivided public good lay beneath all official thinking on radio and television until the 1970s. In 1977 . . .' (Curran and Seaton 1981 p. 310).

4 See, for example, Rex Cathcart's (1984) account of the BBC in Northern Ireland for evidence that it was commercial television and not the BBC which first provided representation of the nationalist community in Northern Ireland and the palpable increase in representation of different ethnic, cultural and linguistic groups in radio via advertising financed stations.

5 The World Bank estimated that there were 409 million English native speakers in the early 1980s. The next largest language community was Hindi/Urdu speakers who numbered 352 million. But the aggregate GNP of the English speakers far exceeded that of the Hindi/Urdu speakers (or any other language community): English speakers' income was US$4,230,375 million whereas Hindi/Urdu speakers' income was US$209,023 million (Wildman and Siwek 1988).

6   Although particularly vulnerable, because of a shared language, to imports from the United States, the world's most powerful information producer (where only aerospace companies export more than Hollywood), the UK has a positive trade balance in film.

7   Accordingly, in 1982, 3,557 English language books were translated into Spanish, but only 816 German language books were translated into Spanish. In the same year, 5,795 English language books and 160 Spanish books were translated into German. And only 108 Spanish books and 873 German books were translated into English in 1982 (UNESCO 1989 p. 336).

8   In 1991 79 per cent of the UK population watched television daily; 62 per cent read a daily newspaper; 47 per cent listened to the radio daily; and 61 per cent attended the cinema at least once annually (Central Statistical Office 1993 pp. 142–4).

9   Potter (1988) has usefully defined five criteria to be satisfied if the consumer's interest is to be adequately served. These are: access, choice, information, redress and representation. To these Sargant (1993) has added a sixth concept: safety.

Chapter 1   Market Forces in Telecommunications

1   Formerly only telephones and other terminals supplied by the monopoly operator could lawfully be used in the UK.

2   International differences in reliability, however, reflect inevitably some differences in methodologies for collecting data. Therefore these comparisons should be treated with a degree of caution.

3   Behind Switzerland, Austria and Italy.

4   BT lags behind New Zealand, France, Sweden and Ireland for digitalization of exchange lines and is the leader, together with Ireland and New Zealand, for trunk digitalization.

5   Electronic exchanges – perhaps a misleading term since the TXE2 and TXE4 switches which filled this category in the early 1970s had a considerable amount of electromechanical components in them – were introduced by the Post Office in the late 1960s and by the mid-1970s accounted for only 17.8 per cent of connections (Cripps and Godley 1978 p. 14). As late as 1988 more than a third of BT's local exchanges were equipped with Strowger electro-mechanical switches which were described by Harper (1989 p. 93) as 'old and largely worn out'.

6   Number of lines.

7   NYNEX claims its subscribers will enjoy a 25 per cent discount on BT's standard charges (*Financial Times* 4–5 March 1995 p. 20).

8   In the 9 kHz to 1 GHz band (partially overlapping the 28–470 MHz band discussed above) defence accounts for 28.8 per cent, the CAA for 3.4 per cent, the BBC for 19.9 per cent, ITV for 18.8 per cent. In the 1–3 GHz band defence accounts for 30.4 per cent of UK spectrum, the CAA for 14.5 per cent, British Telecom for 20 per cent and broadcasting for 11.7 per cent. And in the 3–30 GHz band defence accounts for 37.8 per cent, British Telecom for 22.3 per cent, Mercury for 9.7 per cent, the CAA for 3.6 per cent and satellite broadcasting for 3.3 per cent (Radiocommunications Agency 1994).

Chapter 2   Essential Facilities, Third Party Access and the Problems of Interconnection

1   This argument applies both to BSkyB for satellite and to BT for cable, should BT enter this market and develop a digital CAS.
2   That is, nobody can be made better off without somebody else being made worse off.
3   These are incremental costs, or the costs sustained to increase output by a given amount.

Chapter 3   Concentration of Ownership

1   The 1993 price war initiated by News International's reduction of the cover price of its newspaper titles.
2   If they were competing on price, they would do best by positioning their products as far apart as possible, in order to dim the effects of competition on price–cost margins.
3   Or the share of audience time methodology used by Shew (1994) for News International, which also found the BBC dominant.
4   See *R.* v. *IBA ex parte Rank PLC*, reported in *The Times* 14 March 1986.
5   The Monopolies and Mergers Commission and successive Secretaries of State have been noticeably pliable in interpreting these provisions. As Liberty (Foley 1995) has pointed out, no national newspaper merger has been blocked by this legislation. In 1981, Secretary of State John Biffen consented to News International's purchase of both *The Times* and the *Sunday Times*, despite evidence that the *Sunday Times* was viable as a going concern. This consent was given in spite of Britain having the most concentrated ownership of newspapers in the European Union. The European Institute for the Media (Sanchez-Tabernero 1993) calculated that in 1990 News International's share of UK domestic media was 10 per cent higher than that of any EU media group other than Ireland's Independent Newspapers and Austria's Mediaprint. Indeed, comparing the share of the top two newspaper companies in EU member states again places the UK third behind Ireland and Austria.
6   Not all information media are advertising financed; those that are have varying levels of reliance on advertising revenue. The fact that media products are in different advertising markets does not diminish the potential influence exercised by owners across the whole range of media. Nor can price paid by final consumers be used as an index for definition of markets. Some media, notably terrestrial broadcasting, are not priced, that is consumers do not pay for them at the point of consumption. The UK media do not comprise a single market either for advertising or for final consumption. Yet, though there is no consensus on how influential the media are individually or collectively, the UK media must be considered as a whole when political influence is assessed.

Chapter 4   Universal Service Obligation in Broadcasting and Telecommunications

1 The Assizes were held in Paris in 1989 and were jointly organized and sponsored by the government of France and the Commission of the European Communities. They, and the action programme they initiated, are often known as the Audio-Visual Eureka.
2 The Australian Telecommunications Corporation Act 1989 defines a series of community service obligations including the provision of a telephone service to 'all people in Australia on an equitable basis' (Subsection 27 [4][a]).
3 Positive programme requirements include 'a sufficient amount of high quality national and international news; a sufficient amount of time to be given to other high quality programmes; a range of regional programmes; the provision of religious and children's programmes; and a diversity requirement that programmes should cater for a wide range of tastes and interest' (ITC 1995b p. 7).
4 We do not include in this entitlement a right to speech, for reasons discussed in chapter 5, though we believe that it is in the public interest for all services, and particularly publicly financed services, to give voice to a wide variety of opinions, experiences and personalities and thereby serve the UK's diverse interests and communities.
5 Provided that appropriate safeguards against anti-competitive practices are in place.
6 The figure is for BT lines in 1994. As BT has 98 per cent of residential connections, and all lines of the new operators are digital, the actual figure is slightly higher than this.

Chapter 5   Freedom of Expression

1 The paper deals mainly with principles rather than the detail of the law, so should be applicable to all parts of the UK. However, there are different laws in England and Wales and in Northern Ireland and Scotland. For clarity of argument these differences are ignored.
2 Independent Committee for the Supervision of Standards of Telephone Information Services.
3 It is not really a question of *whether* there is evidence of harm caused to women and children by the use of pornography. There is clearly an abundance of evidence of harm: different kinds of evidence and different kinds of harm, evidence which is consistently concurrent and corroborative. The issue is how to make the harm visible and how to make it matter: how to make women matter (Itzin 1994 p. 12).
4 Expressed by Graham Murdock (at the IPPR Seminar on Expression and Censorship, June 1995). See also Howitt and Cumberbatch (1990).
5 The macro-level statistical study of nations may shadow micro-level differences between individuals and small groups. Therefore the study does not necessarily refute any studies conducted with individuals or small groups, but states quite clearly that there is no general relationship between television violence and violent crimes at a macro level (Wiio 1995).

6 Clive Ponting's revelations about the sinking of the *Belgrano*, the seizure and suppression of a BBC television programme about the *Zircon* spy satellite, or attempts to ban the publication of *Spycatcher*.

7 The Bishop of London's Group on Blasphemy (1988 p. 4). However, since the Rushdie *fatwa* some members of the Group have changed their minds.

8 CEDA, General Recommendation no. 19, GAOR, 47th Session, supp no. 38, 1992, comment on Articles 2(f), 5 and 10(c), para. 12, quoted in Liberty (1994).

9 At the IPPR Seminar on Expression and Censorship, June 1995.

10 Henceforth referred to for convenience as the Lord Chancellor's Department's paper.

11 Personal information was defined 'in terms of an individual's personal life, that is to say, those aspects of life which reasonable members of society would respect as being such that an individual is ordinarily entitled to keep them to himself, whether or not they relate to his mind or body, to his home, to his family, to other personal relationships, or to his correspondence or documents' (cited in Lord Chancellor's Department and the Scottish Office 1993 p. 24).

12 Some may argue that the *News of the World*'s report was justified because teachers are public figures. It seems to us that such a judgement would define 'public figure' too inclusively. Definition of what constitutes a public figure would be central to the implementation of a right to privacy which did not compromise the freedom of the press and the public right to information relevant to full participation in social and political life. Hammering out such a definition could usefully be undertaken by Ofcom through open debate and consultation with media interests, independent experts and the public at large.

13 *Prima facie*, there are grounds for more comprehensive reforms – notably to the laws of libel. Barendt argues that 'the law of libel probably constitutes the most significant legal restriction' on press freedom and freedom of speech (1991 p. 68). Indeed, *New York Times* v. *Sullivan* is exemplary here too. The case was, of course, a libel action and the Supreme Court's decision underscores the United States' commitment to press freedom by emphasizing the importance of not chilling public comment on the conduct of public officials and personalities.

Chapter 6   Audio-Visual Policies: Too Much or Not Enough?

1 News, sports events, games, advertising and teletext are excluded in the computing of transmission time for quota purposes.

2 Such a compromise solution is opposed by the European Parliament, especially on the left. The European Parliament Committee on Culture and Media voted (on 18 January 1996) a revision of the draft proposing, among other measures, a strengthening of the quota 'by appropriate and legally effective means' and its extension to new services, including video-on-demand.

3 This figure refers to 'mostly British' films in 1993. Co-productions had an average budget of £2.32 million. Partial 1994 data on thirty-six out of forty-four UK pictures (National Heritage Committee 1995 vol. I annex F table 7) show an average budget of £3.21 million for 'all British' films. If all

productions with a UK majority share, wherever filmed, are included, the average budget increases to £3.91 million.
4 The Eady Levy, which returned a share of box office revenue to the producers of British films, was scrapped in 1985.
5 UIP, Warner Distributors, Columbia, Buena Vista, Fox.

Chapter 7   Public Service Broadcasting: a Better BBC

1 See Paul Chadwick's salutary *Media Mates* (1989) for an account of the political role exerted by media owners in Australia.
2 Source: BBC internal research (unpublished).
3 At the time of writing the privatization of the BBC transmission structure has been proposed. The estimated value of the sale, £70 million, is to be reinvested in digital infrastructure.
4 Quango (quasi-autonomous non-governmental organization): a body independent from the department that created it, save for the appointment of its members from the ranks of the great and good, nominally under the control of its minister, though not fully accountable to Parliament (*Brewer's Political Dictionary*, quoted in *The Daily Telegraph*, 26 October 1994).
5 As is the CBC in Canada placed under the regulatory authority of the Canadian Radio-television and Telecommunications Commission.
6 Sir Christopher Bland starting from March 1996.

Chapter 8   Convergence and Change: Reforming the Regulators

1 See Curran's discussion paper *Policy for the Press* (1995) written for the IPPR Media and Communication Programme for a well grounded version of this critique.
2 How many cookery programmes are there about preparing novel and impressive dinner parties and how many about cooking for survival and pleasure on a low income?
3 See *The Times* law report of 5 February 1973 (cited in Smith 1974 pp. 169–79).
4 That there have been very few occasions on which the PCC and its predecessors' authority was flouted – a former member estimates fewer than ten times – may be interpreted as evidence either of the regulator's lack of severity, or of the obedience of regulated firms.
5 The Chairman of the PCC, Lord Wakeham at the time of writing, is appointed by the Press Standards Board of Finance (Pressbof) which has ten directors nominated by the newspaper and periodical publishing trade associations. The Chairman of the PCC, the Chairman of Pressbof and three other persons nominated by the Chairman of the PCC, together nominate the Public Members (eight members) and the Press Members (seven members) of the PCC. The Public Members must not be connected with newspaper and periodical publishing and the Press Members must be 'experienced at senior editorial level' (PCC information pack undated). Control, albeit indirect

control, thus rests with the trade associations which finance and control Pressbof.

6 See, *inter alia*, Carsberg (1995); National Consumer Council (1995); the Shadow Chancellor Gordon Brown's, advocacy of a new Competition and Consumer Standards Office; Lord Borrie and Professor Melody at IPPR's Seminar on Law and Regulatory Issues on the Superhighway in September 1995 (Collins 1996).

7 Advertising Standards Authority, British Board of Film Classification, British Broadcasting Corporation, Broadcasting Complaints Commission, Broadcasting Standards Council, Independent Television Commission, Independent Committee for the Supervision of Standards of Telephone Information Services, Press Complaints Commission, Radio Authority, Welsh Fourth Channel Authority.

8 Excepting the forthcoming merger of the Broadcasting Complaints Commission and the Broadcasting Standards Council.

9 For contrary views see the statements by Professor Robert Pinker, Privacy Commissioner of the PCC and formerly a member of the ASA, and Dr Michael Redley, Secretary of the ITC, at the IPPR Seminar on Law and Regulatory Issues on the Superhighway, September 1995 (Collins 1996).

10 The NCC proposes the Gas Consumer Council as a model.

11 Contrast with the Canadian Broadcasting Act 1968, which defined broadcasting as 'any radiocommunications in which the transmissions are intended for direct reception by the general public' (Section 1.3.a).

12 Notably in the recommendations of the Sykes (1923), Crawford (1926), Beveridge (1951) and Annan (1977) Committees of enquiry into broadcasting, and in particular in the Beveridge Committee's proposal for a Public Representation Service.

Appendix 1   The Telecommunications Sector

1 The UK mobile market initially had a strict separation of service providers and network operators, in the sense that a firm could not enter both segments of the market. This rule was eventually relaxed, leading to rapid concentration in the service provider market.

2 Differences in national/urban geography can make FTTH economically viable in some countries/areas but not in others.

3 Digital switches were not cheaper initially, but the price of microprocessors dropped sharply. A country which is now building a telephone network from scratch will face substantially lower costs than those prevailing twenty to thirty years ago.

4 The study was designed to see if there were barriers to customer choice in cabled areas. It highlighted four barriers: the lack of independent and unbiased information on performance of different operators; the lack of information comparing the prices of different operators (whereas cheaper prices were by far the most important factor for persuading users to switch operator); anti-competitive practices engaged in by both BT and cable operators (e.g. inaccurate information about rival operators); the lack of number portability.

5 Among non-cable residential users, 55 per cent said they were happy with

their existing telephone supplier, but 28 per cent said it would have been 'too much hassle' to change. Among business users, 25 per cent were happy with BT and 5 per cent mentioned the change of their advertised number.

6 For historic reasons, local fixed telecommunications services in Kingston upon Hull are provided by Kingston Communications rather than BT, with BT and Mercury providing the alternative options for long distance.

7 Mercury quickly achieved a 50 per cent share in Hull, when the national average was 10 per cent (Department of Trade and Industry 1990).

8 Operators had been more successful in getting their message across to residential users where only 8 per cent of homes in cable areas were unaware of cable telephony, although 39 per cent could not name their local cable company.

References

Adamson, M and D Toole (1995) *Multimedia in the Home*. London. Financial Times Management Reports.

Aitman, D. (1994) Cross ownership of media interests in the European Community. Speech to AIC Conference on Multimedia and Broadcasting Reform, 29–30 September.

Alleman, J (1995) Strategic alliances and a 'common carrier' substitute. *Intermedia* vol. 23 no. 2 pp. 37–9.

Annan Committee (1977) *Report of the Committee on the Future of Broadcasting*. Cmnd. 6753. London. HMSO.

Armstrong, M, S Cowan and J Vickers (1994) *Regulatory Reform: Economic Analysis and the British Experience*. Cambridge, Mass. MIT Press.

Ascherson, N (1978) Newspapers and internal democracy. In Curran, J (ed.) *The British Press: a Manifesto*. London. Macmillan.

Assises européennes de l'audiovisuel (1989) *Project Eureka Audiovisuel*. Paris. Ministère des affaires étrangères république française and Commission of the European Communities.

Baer, W (1994) Telecommunications infrastructure competition: the costs of delay. PICT Policy Research Paper no. 31 (based on EAC/RAND Seminar, November 1994). Mimeo. Brunel University.

Baldwin, R (ed.) (1995) *Regulation in Question: the Growing Agenda*. Report sponsored by Merck, Sharp and Dohme by Dr Baldwin of the Department of Law, London School of Economics and Political Science.

Bangemann, M (1994) *Europe and the Global Information Society: Recommendations to the European Council*. Brussels.

Barendt, E (1991) Press and broadcasting freedom: does anyone have any rights to free speech? *Current Legal Problems* vol. 44 pp. 63–82.

Barendt, E (1994) Legal aspects of charter renewal. *Political Quarterly* vol. 65 no. 1. pp. 20–8.

Barnett, S (1995) The yawning of a new age. *Guardian* 14 September.

Baumol, W and J Sidak (1994) *Towards Competition in Local Telephony*. Cambridge, Mass. MIT Press.

Baxt, R (1992) Regulation: structure and issues. *Media Information Australia* no. 63 February pp. 13–18.

BBC (1993) *An Accountable BBC*. London. British Broadcasting Corporation.

BBC (1994a) *Report and Accounts 1993/94*. London. British Broadcasting Corporation.

BBC (1994b) *The BBC's Fair Trading Commitment*. London. British Broadcasting Corporation.

Bell, D (1973) *The Coming of the Post Industrial Society*. Harmondsworth. Penguin.

Berlin, I (1969) *Four Essays on Liberty*. Oxford. Oxford University Press.

Beveridge Committee (1951) *Report of the Broadcasting Committee 1949*. Cmd. 8116. London. HMSO.

Bishop of London's Group on Blasphemy (1988) *Offences against Religion and Public Worship*. GS Misc. 286. London.

Blumler, J (1986) Television in the United States: funding sources and programming consequences. In Home Office *Research on the Range and Quality of Broadcasting Services*. London. HMSO.

Board of Deputies of British Jews (1992) *Group Defamation: Report of a Working Party of the Law, Parliamentary and General Purposes Committee*. Chaired by Eldred Tabachnik QC. London.

Briggs, A (1961) *The History of Broadcasting in the United Kingdom* vol. 1. Oxford.

British Film Commission (1994) *Annual Report*. London. British Film Commission.

British Film Institute (1995) *Film and Television Handbook*. London. British Film Institute.

British Media Industry Group (1995) *A New Approach to Cross-Media Ownership*. Submission to the Department of National Heritage. London. BMIG.

British Telecom (1995) *A Framework for Effective Competition: BT's Response to Oftel's Consultative Document of December 1994*. London. British Telecom.

Brittan, S (1987) The fight for freedom in broadcasting. *Political Quarterly* vol. 58 no. 1 pp. 3–23.

Brooks, S (1992) Broadcasting regulation in the new era. *Media Information Australia* no. 63 February pp. 9–12.

Bureau of Transport and Communications Economics (1994) *Communications Futures: Final Report*. Chapter 4: Social issues and the question of universal service obligations. BTCE Report 89. Canberra. Australian Government Publishing Service.

Cain, P and J M MacDonald (1991) Telephone pricing structure: the effects on universal service. *Journal of Regulatory Economics* vol. 3 no. 4 pp. 293–308.

Calcutt, Sir D (Chair) (1990) *Report of the Committee on Privacy and Related Matters*. Cmnd. 1102. London. HMSO.

Calcutt, Sir D (1993) *Review of Press Self-Regulation*. Cmnd. 2135. London. HMSO.

Carsberg, B (1995) Need for unitary competition authority. *Financial Times* 24 February p. 15.

Cathcart, R (1984) *The Most Contrary Region*. Belfast. Blackstaff.

Cave, M, C Milne and M Scanlan (1994) *Meeting Universal Service Obligations in a Competitive Telecommunications Sector*. Report to DG IV of the Commission of the European Communities. HCM/UK/5. March.

Central Statistical Office (1993) *Social Trends*. London. HMSO.

Chadwick, P (1989) *Media Mates: Carving Up Australia's Media*. South Melbourne. Macmillan.

Chancellor of the Duchy of Lancaster (1993) *Open Government.* Cmnd. 2290. London. HMSO.

Chippindale, P and C Horrie (1988) *Disaster! The Rise and Fall of News on Sunday.* London. Sphere.

Collins, R (ed.) (1996) *Converging Media: Converging Regulation?* London. IPPR.

Collins, R and J Purnell (eds) (1996) *Reservoirs of Dogma.* London. IPPR.

Commission of the European Communities (1987) *Towards a Dynamic European Economy.* Green Paper on the development of a common market for telecommunication services and equipment. COM (87) final. Brussels. Commission of the European Communities.

Commission of the European Communities (1993) White Paper on *Growth Competitiveness and Employment.* COM (93) 700 final. Brussels. Commission of the European Communities.

Commission of the European Communities (1994a) Green Paper on *Audiovisual Policy in the European Union. Strategy Options to Strengthen the European Programme Industry in the Context of the Audiovisual Policy of the European Union.* COM (94) 96 final. Brussels. Commission of the European Communities.

Commission of the European Communities (1994b) Green Paper on the *Liberalisation of Telecommunications Infrastructure and Cable Television Networks.* Part One. COM (94) 440 final. Brussels. Commission of the European Communities.

Commission of the European Communities (1995) Green Paper on the *Liberalisation of Telecommunications Infrastructure and Cable Television Networks.* Part Two. COM (94) 682 final. Brussels. Commission of the European Communities.

Consumers' Association (1991) Broadcasting in the 1990s: the policy issues from a consumer perspective. Mimeo. London.

Consumers' Association (1993) *The Future of the BBC: Response to the Department of National Heritage Consultation.* London. Consumers' Association.

Corry, D, D Souter and M Waterson (1994) *Regulating Our Utilities.* London. IPPR.

Council of the European Communities (1989) Directive on the coordination of certain provisions laid down by law regulation or administrative action in Member States concerning the pursuit of television broadcasting activities (Television Without Frontiers Directive). 89/552/EEC. OJL 298. 17 October.

Council of the European Communities (1995) Directive on the use of standards for the transmission of television signals. 95/47/EC. OJL 281. 23 October.

Crawford Committee (1926) *Report of the Broadcasting Committee 1925.* Cmd. 2599. London. HMSO.

Cripps, F and W Godley (1978) *The Planning of Telecommunications in the United Kingdom.* Cambridge. Department of Applied Economics, University of Cambridge.

Curran, J (1995) *Policy for the Press.* London. IPPR.

Curran, J and J Seaton (1981) *Power without Responsibility.* London. Fontana.

Curran, J and J Seaton (1991) *Power without Responsibility.* London. Routledge.

Davies, M (1995) The New Zealand experience. Paper presented at 1995 International Workshop on Interconnection. Abstract from web page.

Department of National Heritage (1994) *The Future of the BBC.* Cmnd. 2621. London. HMSO.

Department of National Heritage (1995) *Media Ownership: the Government's Proposals*. Cmnd. 2872. London. HMSO.

Department of Trade and Industry (1982) *The Future of Telecommunications in Britain*. Cmnd. 8610. London. HMSO.

Department of Trade and Industry (1992) *Abuse of Market Power*. Cmnd. 2100. London. HMSO.

Department of Trade and Industry (1994) *Study of the International Competitiveness of the UK Telecommunications Infrastructure*. London. PA Consulting Group.

Dertouzos, J and W Trautman (1990) Economic effects of media concentration: estimates from a model of the newspaper firm. *Journal of Industrial Economics* vol. XXXIX no. 1 pp. 1–14.

Director-General of Fair Trading (1986) *Annual Report of the Director General of Fair Trading 1985*. London. HMSO.

Durlacher (1995) *A Survey of the Video and Computer Games Industry*. London. Durlacher.

Dworkin, A (1993) *Letters from a War Zone*. Brooklyn. Lawrence Hill Books.

Foley, C (ed.) (1995) *Human Rights, Human Wrongs: the Alternative Report to the United Nations Human Rights Committee*. Liberty. London. Rivers Oram Press.

Garnham, N (1993) Public service broadcasting and the consumer. *Consumer Policy Review* July pp. 152–8.

Garnham, N (1994) The broadcasting market. *Political Quarterly* vol. 65 no. 1 pp. 11–19.

Geroski, P (1991) *Market Dynamics and Entry*. London. Blackwell.

Gibbons, T (1991) *Regulating the Media*. London. Sweet & Maxwell.

Glasgow University Media Group (1976) *Bad News*. London. Routledge & Kegan Paul.

Glasgow University Media Group (1980) *More Bad News*. London. Routledge & Kegan Paul.

Gore, A (1994) Plugged into the world's knowledge. *Financial Times* 19 September p. 22.

Graham, A and G Davies (1992) The public funding of broadcasting. In Congdon, et al. (eds) *Paying for Broadcasting: the Handbook*. London. Routledge.

Gray, L (1992) Ownership and control: some thoughts for the future. *Media Information Australia* no. 63 pp. 19–22.

Hall, S (1978) Newspapers, parties and classes. In Curran, J (ed.) *The British Press: a Manifesto*. London. Macmillan. pp. 29–52.

Harper, J (1989) *Telecommunications Policy and Management*. London. Pinter.

Hodgson, P (1992) Foreword. In Congdon, T et al. (eds) *Paying for Broadcasting: the Handbook*. London. Routledge.

Home Office (1979) *Report of the Committee on Obscenity and Film Censorship*. Cmnd. 7772. London: HMSO.

Home Office (1988) *Broadcasting in the '90s: Competition, Choice and Quality*. Cmnd. 517. London. HMSO.

House of Commons Library (1994) *Digital and High Definition Television*. Research Paper 94/83. London. House of Commons.

Howitt, D and G Cumberbatch (1990) *Pornography: Impacts and Influences*. Commissioned by the Home Office Research and Planning Unit.

ICSTIS (1993) *Premium Rate Telephone Services: Consumers' Experience and Attitudes*. London. ICSTIS.

ICSTIS (1995) *Activity Report 1994.* London. ICSTIS.

Independent Broadcasting Authority (1974) *Evidence to the Committee on the Future of Broadcasting* paragraphs 205–6. London. IBA.

IPPR (1991) *The Constitution of the United Kingdom.* London. Institute for Public Policy Research.

IPPR (1993) *Social Justice in a Changing World.* London.

IPPR (1994) *Report of the Commission on Social Justice.* London. Institute for Public Policy Research.

ITC (1994) ITC response to BBC White Paper. Press release. 5 November. London. Independent Television Commission.

ITC (1995a) ITC to draw up code on conditional access for subscription television. Note to editors. Press Release. 18 July. London. Independent Television Commission.

Itzin, C (1994) Pornography as hate speech: inciting hatred and violence against women. Paper presented as part of series of Seminars on Aspects of Women's Human Rights, 11 May. Department of Law, School of Oriental and African Studies, University of London.

Jong, E (1996) Deliberate lewdness and the creative imagination: should we censor pornography? In Collins, R and J Purnell (eds) *Reservoirs of dogma.* London. IPPR. pp. 10–24.

Kay, J (1995) Presentation at Oftel's Public Hearing 23 November. Mimeo.

Labour Party (1974) *The People and the Media.* London. Labour Party.

Labour Party (1991) *Arts and Media: Our Cultural Future.* London. Labour Party.

Labour Party (1994) *Winning for Britain.* London. Labour Party.

Lee, S (1990) *The Costs of Free Speech.* London. Faber and Faber.

Liberty (1994) *Censored: Freedom of Expression and Human Rights.* Report no. 8 in the Human Rights Convention Series. London. National Council for Civil Liberties.

London Economics (1994a) *Barriers to Entry and Exit in UK Competition Policy.* Office of Fair Trading Research Paper no. 2. London. London Economics.

London Economics (1994b) *The Future of Postal Services.* London. London Economics.

London Economics (1994c) *The Economic Impact of Television Quotas in the European Union.* London. London Economics.

London Economics and BIPE Conseil (1994) *White Book of the European Exhibition Industry.* Milan. Media Salles.

Lord Chancellor's Department and the Scottish Office (1993) *Infringement of Privacy.* CHAN J06091 5NJ 7/93. London. Central Office of Information.

Lovell, T (1980) *Pictures of Reality.* London. British Film Institute.

McGregor, O (Chair) (1977) *Report of the Royal Commission on the Press.* Cmnd. 6810. London. HMSO.

McKay, R and B Barr (1976) *The Story of the Scottish Daily News.* Edinburgh. Canongate.

McLuhan, M (1962) *The Gutenberg Galaxy.* London. Routledge & Kegan Paul.

Melody, W (1980) Radio spectrum allocation: role of the market. *The American Economic Review* vol. 70 no. 2 pp. 393–7.

Melody, W (1990) *Telecommunications Policy Directions for Australia in the Global Information Economy.* Melbourne. CIRCIT.

Mitchell, J (1994) Note to the Department of National Heritage on behalf of the Voice of the Listener and Viewer Consumer Broadcasting Liaison Group: Broadcasting Complaints.

Mitchell, J and J Blumler (1994). Television and the Viewer Interest. John Libbey. London.

Mitchell, J and C Milne (1990) *Quality of Service Indicators for the Telephone Service to Residential Consumers.* CCIS Policy Paper no. 1. London. Centre for Communication and Information Studies, Polytechnic of Central London.

Modood, T (1994) Establishment, multiculturalism and secularism. *Political Quarterly* vol. 65 no. 1. pp. 53–73.

Murdoch, R. (1989) Freedom in Broadcasting. The McTaggart Lecture. Edinburgh Television Festival. 25 September.

National Consumer Council (1994) *A Broadcasting Consumer Council Response to The Consultation Document 'The Future of the BBC'.* 6 July. London. National Consumer Council.

National Consumer Council (1995) *Competition and Consumers.* London. National Consumer Council.

National Heritage Committee (1995) *The British Film Industry, Second Report, Session 1994–5* vols I, II, III. London. HMSO.

Negroponte, N (1995) *Being Digital.* London. Hodder & Stoughton.

NERA (1992) *Market Definition in UK Competition Policy.* Office of Fair Trading Research Paper no. 1. London. National Economic Research Association.

Neuman, R (1991) *The Future of the Mass Audience.* Cambridge. Cambridge University Press.

New Media Markets (1995) Three of ten firms not aware of cable telephony. *New Media Markets* vol. 13 no. 44. 7 December. pp. 10–11. London. *Financial Times.*

Noam, E (1987) The public telecommunication network: a concept in transition. *Journal of Communication* vol. 37 no. 1 pp. 28–48.

Nora, S and A Minc (1978) *The Computerization of Society.* Cambridge. MIT Press.

OECD (1990) *Performance Indicators for Public Telecommunications Services. Report by the Working Party on Telecommunications and Information Services.* Paris. Organization for Economic Co-operation and Development.

OECD (1993) Communications Outlook 1993. Paris. Organization for Economic Co-operation and Development.

Office of Science and Technology (1995a) *IT and Electronics.* Technology Foresight Progress through Partnership report no. 8. London. Department of Trade and Industry.

Office of Science and Technology (1995b) *Leisure and Learning.* Technology Foresight Progress through Partnership report no. 14. London. Department of Trade and Industry.

Oftel (1994a) *A Framework for Effective Competition* December. London. Oftel.

Oftel (1994b) *Market Report.* London. Oftel.

O'Malley, T (1994) *Closedown: the BBC and Government Broadcasting Policy 1979–92.* London. Pluto.

O'Malley, T and J Traherne (undated) *Selling the Beeb?* (1993). London. Campaign for Press and Broadcasting Freedom.

Peacock, A (Chair) (1986) *Report of the Committee on Financing the BBC.* Cmnd. 9824. London. HMSO.

Pilkington, H (Chair) (1962) *Report of the Committee on Broadcasting 1960.* Cmnd. 1753. London. HMSO.

Policy Studies Institute (1995) *Barriers to Telephone Universal Service*. London. Policy Studies Institute.

Porat, M (1977) *The Information Economy: Definition and Measurement*. US Department of Commerce, Office of Telecommunications. May. OT 77–12 (1). Washington DC. US Government Printing Office.

Potter J (1988) Consumerism and the public sector: how well does the coat fit? *Public Administration* vol. 66 no. 2 pp. 149–64.

Public Network Europe (1995) *1995 Yearbook*. London. The Economist.

Radiocommunications Agency (1993) *Annual Report and Accounts 1992–93*. London. HMSO.

Radiocommunications Agency (1994) *The Future Management of the Radio Spectrum*. London. Radiocommunications Agency.

Robinson, C (1993) *But Who Will Regulate the Regulators?* London. Adam Smith Institute.

Rowlands, C. (1994) Joint ventures: a blueprint for the future. Speech to AIC Conference on Multimedia and Broadcast Reform, 29–30 September.

Rushdie, S (1992) *The Satanic Verses* (1988). Dover. The Consortium.

Sanchez-Tabernero, A (1993) *Media Concentration in Europe: Commercial Enterprise and the Public Interest*. Media Monograph no. 16. European Institute for the Media, Dusseldorf.

Sargant, N (1993) Listening to the consumer. *Consumer Policy Review* vol. 3 no. 3 pp. 159–66.

Scannell, P (1989) Public service broadcasting and modern public life. *Media Culture and Society* vol. 11 no. 2 pp. 135–66.

Scharf, A (1994) Open letter to Mr Silvio Berlusconi, President of the Council, from Mr Albert Scharf, President of the European Broadcasting Union (EBU). Press release PR 11/94. Geneva. European Broadcasting Union.

Schlesinger, P (1987) *Putting Reality Together*. London. Methuen.

Senate (1992) *Broadcasting Services Bill 1992. Explanatory Memorandum*. Canberra. The Parliament of the Commonwealth of Australia.

Shew, W (1994) *UK Media Concentration*. Prepared for News International PLC. July 1994. London. Arthur Andersen Economic Consulting.

Short, C (1996) Page Three, incident display and the roots of concern. In Collins, R and J Purnell (eds) *Reservoirs of Dogma*. pp. 33–41. London. IPPR.

Smith, A (1972) Television coverage of Northern Ireland. *Index on Censorship* vol. 1 no. 2 pp. 15–32.

Smith, A (1974) British Broadcasting. Newton Abbot. David and Charles.

Smith, A (1978) The audience as tyrant. In *The Politics of Information: Problems of Policy in Modern Media*. London. Macmillan.

Society of Telecom Executives (1983) Liberalisation, privatisation and regulation: what future for British Telecom? Mimeo.

Stewart, J, L Kendall and A Coote (1994) *Citizens' Juries*. London. IPPR.

Sykes Committee (1923) *Broadcasting Committee Report*. Cmd. 1951. London. HMSO.

Taylor, C (1993) *Reconciling the Solitudes: Essays on Canadian Federalism and Nationalism*. Montreal and Kingston. McGill-Queen's University Press.

Temple Lang, J (1994) Defining legitimate competition: companies' duties to supply competitors and access to essential facilities. *Fordham International Law Journal* vol. 18 pp. 437–524.

The Economist (1994a) The future of the BBC. *The Economist* 9 July.

The Economist (1994b) A yearning for direct democracy. *The Economist* 9 July.

Touche Ross (1993) *Setting the Level of the Television Licence Fee.* A study for the Department of National Heritage. London. HMSO.

Trevelyan, J (1973) *What the Censor Saw.* London. Michael Joseph.

UNESCO (1989) *World Communication Report.* Paris. UNESCO.

Veljanovski, C (1987) *Commercial Broadcasting in the UK: Over-regulation and Misregulation.* London. Centre for Economic Policy Research.

Veljanovski, C and W Bishop (1983) *Choice by Cable.* London. Institute of Economic Affairs.

Wall, D and J Bradshaw (1994) The Message of the Medium. *New Law Journal* 9 September pp. 1198–9.

Wall, S (1977) The UK: an information economy? Mimeo. British Telecom.

Wiio, O (1995) Is television a killer? An international comparison. *Intermedia* vol. 23 no. 2 pp. 26–31.

Wildman, S and S Siwek (1988) *International Trade in Films and Television Programs.* Cambridge. Ballinger.

Williams, B (1996) Suppression and restriction of pornography: long spoons and the supper of the righteous. In Collins, R and J Purnell (eds) *Reservoirs of Dogma.* London. IPPR. pp. 25–33.

Glossary

analogue   The method of transmission where the information is turned into a wave, with the peaks and troughs representing various attributes of the message (e.g. for a voice these might be volume and tone). See **digital**.

any to any   A description of a network where any subscriber to the network is able to communicate with any other subscriber or service provider on the network.

bandwidth   Describes the capacity of a communications channel. For example full motion video traffic requires a higher bandwidth than voice. See **full motion video**.

broadcast   A mode of communication where a party communicates one way with a large number of other parties. Each receiving party receives the same signal. The main examples are television and radio. See **one way**. Also known as one to many or point to multipoint systems.

cable network   The name given to a network constructed for the local distribution of television systems by wire (coaxial or fibre). In some countries, such as the UK, cable licensees are also allowed to build local switched telecoms networks. Such networks are sometimes called CATV networks. See **DTH, fibre optic cable, terrestrial**.

CATV   See **cable network**.

CD-ROM   A compact disk based computer storage technology able to store very large quantities of data, making new types of application possible, such as very graphical games and electronic encyclopaedias.

client   A computer accessing or subscribing to an on-line network. See **host**.

closed user group   A group with access to communication and information services which are inaccessible to non-members of the closed user group.

common carrier   An operator that carries communication from any source (such as a PTO or an Internet service provider) rather than an operator which keeps control over the content carried (such as a broadcaster). See **Internet service provider.**

compression   The ability to reduce the bandwidth that a given signal requires, thus increasing the number of services that can be carried by a given amount of bandwidth. This is usually achieved by removing the unnecessary components of the signal. See **bandwidth.**

conditional access   A CA system encrypts a signal so that it cannot be intercepted by unauthorized users, such as defaulted subscribers. Typically a set-top box decrypts the signal. See **encryption, set-top box.**

connectivity   A measure of the degree to which users are able to communicate or connect with each other. An any to any network is a high connectivity network. Unconnected users are often said to desire connectivity. See **any to any.**

content   The actual information or message making up telecoms traffic. This could be for example a film, a voice conversation, a written story or a photograph. Content has an equivalent wider meaning beyond the telecoms environment.

customer premises equipment   The locally located unit(s) used by the subscriber to connect to a given network (e.g. telephone, television set, computer or modem).

digital   The method of transmission where the information is turned into a series of 1s and 0s. Any carrier wave has, as a result, only an on and an off setting. Among other attributes, this greatly reduces signal degradation and allows perfect replication of any stored information.

digital compression   Electronic technique for reducing the number of binary digits necessary to convey information. In broadcasting, this reduces the amount of information which needs to be transmitted in order to provide acceptable sound and picture quality.

distribution network   Any network over which information is transmitted to users. Examples include a phone network or a cable TV network.

DTH   In direct to home (DTH) television networks, the signal is broadcast directly from a geostationary satellite to individual homes via local satellite antennas (i.e. dishes). See **cable network, terrestrial.**

e-mail   Messages, typically text, carried electronically within an office environment or over the Internet. See **Internet.**

encryption   The disguising of a signal so that it cannot be intercepted by unauthorized users. The term is most often used in relation to the DTH pay TV market. See **conditional access, set-top box.**

fibre optic cable   Cable made up of very small glass filaments. Signals are transmitted as bursts of light. Fibre cables have an extremely high bandwidth.

fragmentation Describes the rapid increase in the number of titles, channels or services within a given medium, such as magazines or television.

free-to-air Television or radio services funded from the TV licence fee or advertising, with no further charge to the viewer or listener.

frequency channel Area on the electromagnetic spectrum used for transmission of broadcast services. Analogue transmission allows one programme service per frequency channel, digital allows a single frequency channel to carry several programme services.

full motion video Simply a video image moving at normal speed, such as television or cinema. Full motion video can now be played out on computers as well as on televisions.

hardware Equipment used in computer and telecommunications systems and networks. See **software.**

host A computer storing files or information which make up part of an on-line service or a computer that administers traffic for an on-line service. Typically these are the same machine. See **client.**

information provider The usual name given to a provider of content to an on-line service. See **content, on-line.**

interactivity The ability of two parties to both send and receive messages. The term is often used loosely and does not imply that the two parties have equal opportunity to send and receive.

interconnection The interlinking of two distinct networks. For example, in the UK BT's and Mercury's networks are interconnected.

Internet A global computer network comprising hosts, where information is stored or processed, and clients, which access the network via a dial-up modem or through a direct connection with a host. The hosts are connected via a series of leased lines.

Internet service provider A company providing individuals or businesses with access to the Internet. More generally, service providers can offer access to services other than the Internet.

ISDN Integrated switched digital networks (ISDNs) allow digital communication at higher bandwidths than are usually associated with the phone network. See **bandwidth.**

local loop The local part of a telecommunications network, connecting the user to the local exchange.

multiplex The combination of several programme services, and possibly additional services, within a frequency channel.

navigation software Software that allows a user to easily select what service to receive. Web browsers are a type of navigation software. It is envisaged that future set-top boxes will require sophisticated navigation software as the number of services available increases. See **set-top box, web browser.**

near-video-on-demand   NVOD is an environment which simulates video-on-demand. Typically refers to a broadcast system where the same piece of programming is shown with a time delay across several channels. Thus, to the consumer it appears that the (e.g.) movie is always available within ten or fifteen minutes. See **video-on-demand.**

one way   A mode of communication where sending parties are unable to receive information back from receiving parties. See **two way.**

on-line   Description of a PC based service whereby many local machines (the clients) interact with a larger distant machine (the host) by using a telecoms network. Typically used to refer to commercial operations (such as Compuserve) and the Internet, rather than intra-office applications.

pay per view   In a PPV system the subscriber pays a separate fee for each programme watched. Usually considered in the context of movies or sports TV channels.

pay TV   Television services that the consumer must pay for in order to receive.

point to multipoint   See **broadcast.**

point to point   A mode of communication where there are two parties to any exchange. The telephone is the best known example.

PSTN   The public switched telephone network (PSTN) is the main telephone network for public (as opposed to closed user group), switched, voice communication. See **closed user group, switched network.**

PTO   A public telecommunications operator managing all or part of a country's PSTN. The term is also sometimes applied to any telecommunications operator. See **PSTN.**

set-top box   The unit that decodes and decrypts a signal so that it can be viewed by the user. Typically a pay TV device, the unit is a small box that sits on or close to the television. See **conditional access, encryption.**

simulcasting   Simultaneous transmission of a programme in analogue and digital form.

software   The code/programmes that instruct hardware how it should function in certain situations. For example, a laptop computer is hardware. It runs because of its operating software (allowing it to perform general functions such as to switch on and control the screen) and its applications software (allowing it to perform specific tasks, such as to write a document). See **hardware.**

switched broadband network   A network allowing two way, point to point communication at bandwidths high enough to allow video traffic. See **bandwidth, point to point, two way.** Also known as a full service network or video dial tone.

switched network A network where signals can be routed efficiently from one party to another.

terrestrial Describes television signals broadcast from a ground based transmission site. See **cable network, DTH.**

two way A mode of communication where all parties are able to send and receive information. See **one way.**

universal access A common regulatory principle whereby any party wishing to be connected to a network should be provided with a connection. Usually referred to as a service obligation because of the received wisdom that it involves connecting geographically distant and hence economically unviable customers.

video-on-demand In a VOD environment a subscriber is able to receive a full motion video signal (e.g. a film) at the exact time desired (i.e. on demand). See **near video on demand.**

web browser A piece of client software able to interpret and display HTML (i.e. World Wide Web) files. The main examples are Netscape and Mosaic.

web site An Internet host storing HTML files or a specific subset of files stored on that host. For example there is a Tottenham Hotspur Supporters site.

wide-screen television Transmission and reception format providing viewers equipped with appropriate receivers with a wide-screen image.

wireless network A network where the user's device (e.g. handset) does not have to be physically connected to the network. Cellular telephony is the most widespread example.

World Wide Web A subset of the Internet. The WWW comprises all that information in the language HTML, and thus recognized by web browsing software. The WWW has enjoyed enormous growth as it is more user-friendly and attractive to look at than other Internet applications. See **web browser, web site.**

Index